IMAGES
of America

TOPSFIELD FAIR
AMERICA'S OLDEST

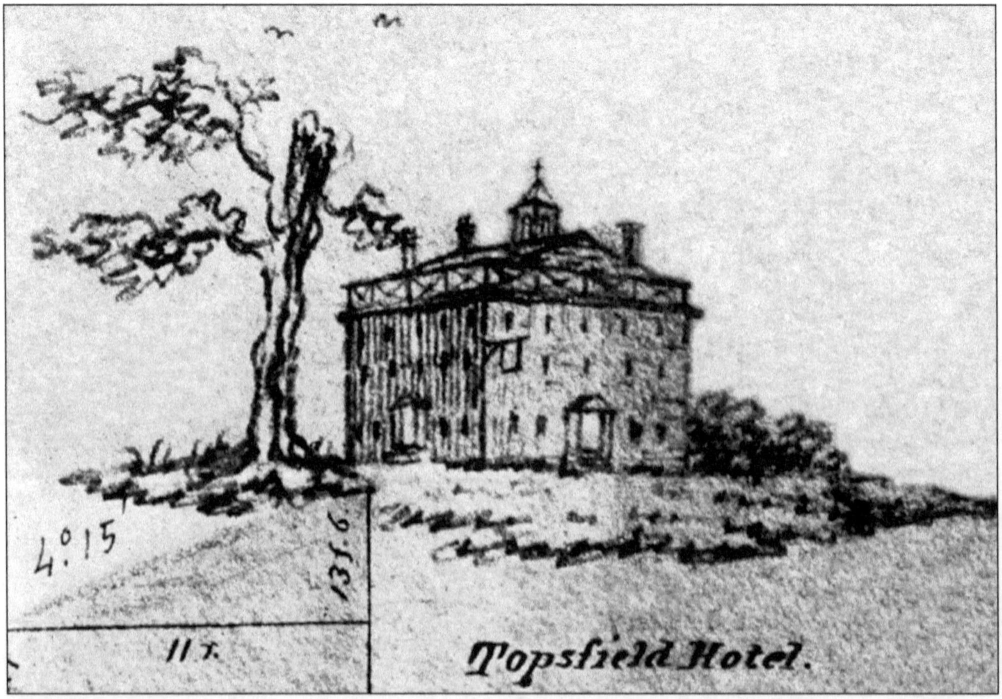

The first meeting was called to organize the Essex Agricultural Society in February 1818 at Cyrus Cummings's tavern in the Topsfield Hotel. The hotel was built at the upper corner of High Street and Route 1 in 1803–1804 by the Newburyport Turnpike Company, which also built the turnpike as a toll road. At that first meeting, Timothy Pickering of Salem was elected president of the newly formed society for the promotion of agriculture. In 1844, the hotel was taken down and removed to Danvers at Clifton Beach and was rebuilt. The business was unsuccessful there, and the building suspiciously burned to the ground in 1846. (Photograph courtesy of the Topsfield Public Library.)

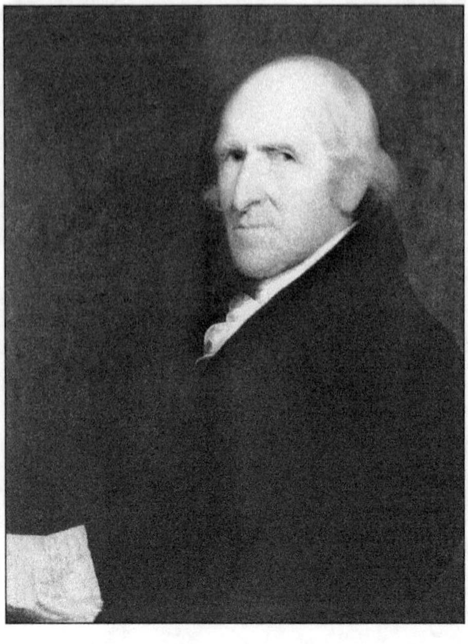

Col. Timothy Pickering of Salem was elected to lead the Essex Agricultural Society on February 18, 1818, at the Topsfield Hotel in Topsfield. Colonel Pickering, a prominent farmer in Massachusetts, was a former member of George Washington's cabinet, serving as secretary of war and postmaster general. Pickering was also a judge, a statesman, and a highly respected member of the community. He led the society as president for 11 years.

IMAGES
of America

TOPSFIELD FAIR
AMERICA'S OLDEST

David H. Fletcher

ARCADIA
PUBLISHING

ISBN 978-1-5316-0804-0

Published by Arcadia Publishing
Charleston, South Carolina

Library of Congress Catalog Card Number: applied for.

For all general information, contact Arcadia Publishing:
Telephone 843-853-2070
Fax 843-853-0044
E-mail sales@arcadiapublishing.com
For customer service and orders:
Toll-free 1-888-313-2665

Visit us on the Internet at www.arcadiapublishing.com.

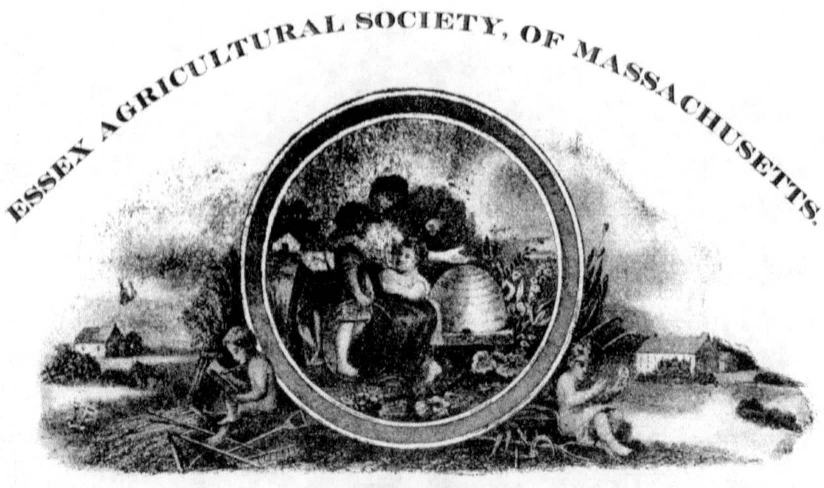

The early logo of the Essex Agricultural Society of Massachusetts was used on all official citations and letterheads. The society was founded in 1818 in Topsfield for the advancement of agriculture and to share the latest developments in farming with others.

Contents

Acknowledgments		6
Introduction		7
1.	The Early Days	9
2.	The Fairgrounds	19
3.	The 1920s and 1930s	25
4.	4-H Activities	55
5.	Special Awards	65
6.	Animals and Demonstrations	71
7.	Entertainment	99
Topsfield Fair Timeline		128

Acknowledgments

I would like to thank all the wonderful and generous people at the fair office and the Essex Agricultural Society, as well as those from around the county, for their help and encouragement in preparing this volume. A special thank-you goes to those people who either loaned or donated their old photographs to the society for various uses.

The dedicated department and show chairmen, along with their respective committees and the exhibitors, are what is needed to make up a "grand fair institution" such as the Topsfield Fair.

For decades, beginning prior to 1930, three generations of Fletchers have been associated with the fair in many capacities and have shown numerous animals in addition to serving on many of the fair's committees.

I am dedicating this volume to one of the most dedicated of those persons, Harold T. Fletcher. My uncle Harold, an enthusiastic poultryman, was associated with the poultry department from the 1930s until his passing in 2002 in many capacities, from exhibitor to show superintendent. He always had "a few birds" to exhibit from his backyard coops in Topsfield, where he resided with his wife, Agnes, for most of his life. To Agnes's chagrin, Harold usually had more birds after the fair than before as a result of a whole week of "horse trading."

Harold and his brother Albert, my father, were both encouraged to raise and show poultry, especially pigeons and bantams, by their father, Harry T. Fletcher, an early resident of Topsfield village. My grandparents lived on Main Street in the very center of town, and my grandfather's flock of Grey Call ducks could often be seen flying back and forth over Main Street to the pond down the street. All three of these men encouraged me to raise and show poultry at various fairs and exhibitions.

Without leaders such as Harold Fletcher and all the other department chairmen, the Topsfield Fair would never have survived for nearly two centuries to become America's oldest agricultural fair.

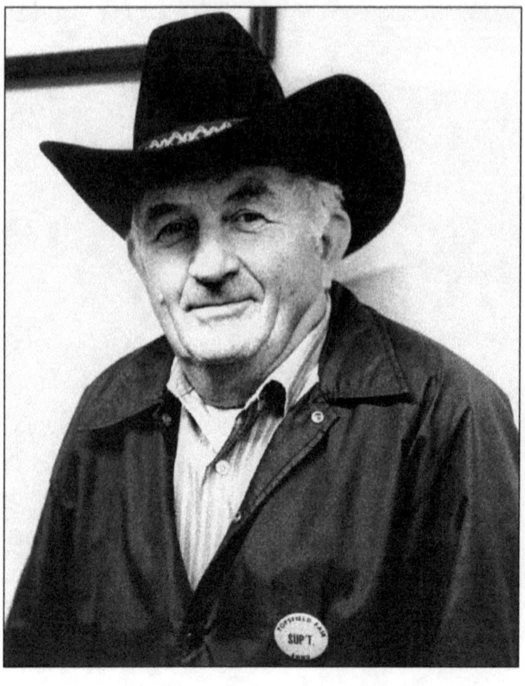

Harold T. Fletcher started showing poultry at the Topsfield Fair in the 1920s at age 15 and went on to become superintendent of the poultry show for decades. He and his two brothers grew up on Main Street in Topsfield, and they all graduated from Topsfield High School. In 2001, Harold and his wife, Agnes, moved from Topsfield to their sons' farm in Fryeburg, Maine, to retire once again. Harold was recognizable around the town and at the fair by the black western-style hat that he usually wore. He had many of the photographs in this book donated to the society's archives before he moved to Maine, where he passed away in 2002. (Photograph courtesy of Agnes Fletcher.)

INTRODUCTION

As an introduction to this book, the author would like to emphasize the significance and wholesome purpose of the Essex Agricultural Society. Can one's mind conceive of the far-reaching effects of the work of the society in the past 185 years? Even with the slow means of travel by ox and horse in the earlier years, farmers learned early the value of getting together for friendly competition and the exchange of ideas and found the annual cattle shows a source of pleasure and profitable education to themselves. Later, through experiments on the Essex Agricultural Society's Treadwell Farm and institutes in Topsfield, the members endeavored to increase their knowledge of agricultural problems. Today, through close cooperation with the various agricultural agencies, the society helps to bring the best of exhibits and demonstrations to all who attend the fair. Down through the years, each set of officers and committees has had different ideas as to how the society could best serve its members and the county, and all have contributed to the influence and benefits that the society as a whole has bestowed on our county. Seldom do we stop to think how great may be the effect of an organization such as this one over almost two centuries.

The need for some type of agricultural organization to be formed in Essex County for getting farmers together and sharing information was recognized at the beginning of the 19th century. In 1818, a notice was posted in the newspapers of the towns in the county that on February 16, 1818, a meeting would be held at Cyrus Cummings's tavern in the Topsfield Hotel on the Newburyport Turnpike (Route 1) in Topsfield to form such an organization. When the meeting was over, Col. Timothy Pickering was the president of the newly formed Essex Agricultural Society.

Just two years later, in 1820, the society held its first cattle show in Topsfield so that all interested persons in the county could get together for sharing agricultural ideas and methods and to have some friendly competition between the farmers in cattle showing, oxen pulls, and the most popular event, the plowing matches. Over the years, these one-day cattle shows gradually turned into fairs in which other agricultural livestock and endeavors were shown. After several years had passed, women were admitted as members, and their home skills and crafts were exhibited. The exhibitions grew to two-day and then three-day fairs.

These annual events were held in many towns all over the county until 1895, when the society decided it needed a central and more permanent location to hold the fair. It purchased 50 acres in Peabody and built a horse racing track and some buildings to house the exhibitions and livestock. The fair was held in Peabody until 1910, when the property was sold and the fair moved to the society's 150-acre Treadwell experimental farm in Topsfield. The farm was bequeathed to the society in 1856 by Dr. John Goodhue Treadwell, but certain conditions attached to the gift made using the site for a fair less than ideal at that time.

Essex County farmers—or agriculturalists, as they are called today—were bold and adventurous in their search for new agricultural endeavors to develop on their farms to make a living. Some of these included early experimentation in the 1820s and 1830s with developing a silk-production industry that, according to society records in the archives, "women, children and the infirm" could operate on the family farm. There were several large mulberry tree orchards as well as numerous small ones planted in the county to feed the silkworms, but several factors, including some severe winters that decimated the trees, finally discouraged the farmers from continuing their culture. One such orchard in Byfield held some 22,000 mulberry trees. Another early endeavor (starting in 1810) was the raising of Marino sheep for their long and superior wool fiber, but this too would prove to be an unsuccessful husbandry in the county for various reasons. Cranberries were also grown on many bogs in the 19th century, including ones in Andover and Topsfield. Some of the Topsfield cranberries were even grown on the society's Treadwell Farm. Again the colder climate of the county proved a bit too severe for reliable production on the farms here. The cranberry bogs were better suited to the Cape Cod area of the state, where we find cranberries grown today.

As urbanization proceeded around the county, members of the public were searching for more excitement than just the livestock exhibitions. To answer this need and to increase their attendance revenues, the society started including vaudeville attractions, eating establishments, games of chance, and even horse racing. Eventually, several types of racing were held at the annual event, including motorcycle and automobile racing. By the 1950s, greyhound dog racing was a very popular event at all the fairs at Topsfield. Between 1921 and 1925, more than 15 major permanent exhibition halls and stables were erected. Most of those buildings are still being used for that purpose along with several modern buildings, such as Coolidge Hall and the livestock arena. Also, many smaller buildings have been added to the grounds.

Mixed among all the concessions and the large midway (features that tend to overwhelm the view of all the buildings), most of the earlier agricultural events and exhibitions are still present. The total numbers of the individual animal, plant, and home arts exhibits may be far smaller than in earlier years, but most types of exhibits are still well represented for all the 500,000-plus fair visitors each year.

Some of the newer and exciting attractions at the Topsfield Fairs of late are the sand sculptures, renewal of the horse shows (including draft horse shows), and the favorite of everyone—the Musical Ride, performed by the Royal Canadian Mounted Police. The fair also hosts tractor pulls of all kinds, world-famous singing and instrumental groups of entertainers, and of course the giant pumpkin contest, which sees larger pumpkins every year. Each year there are new experiences to be had by all who attend the "oldest agricultural fair in America."

After all these many years, the Topsfield Fair is very unique in many ways. It speaks of agriculture, education, camaraderie, competition, family fun, and past accomplishments and future possibilities. Somehow, the fair combines all of these qualities and more into a package that fair patrons fondly refer to as an "experience." Through the years, the dedication of founders, directors, trustees, staff members, volunteers, and exhibitors has been the lifeblood of the Topsfield Fair. As we move into the 21st century, the fair's continuing success will be predicated on and directly related to the ongoing support of those who are willing to be involved and who are entirely committed to its mission.

One

The Early Days

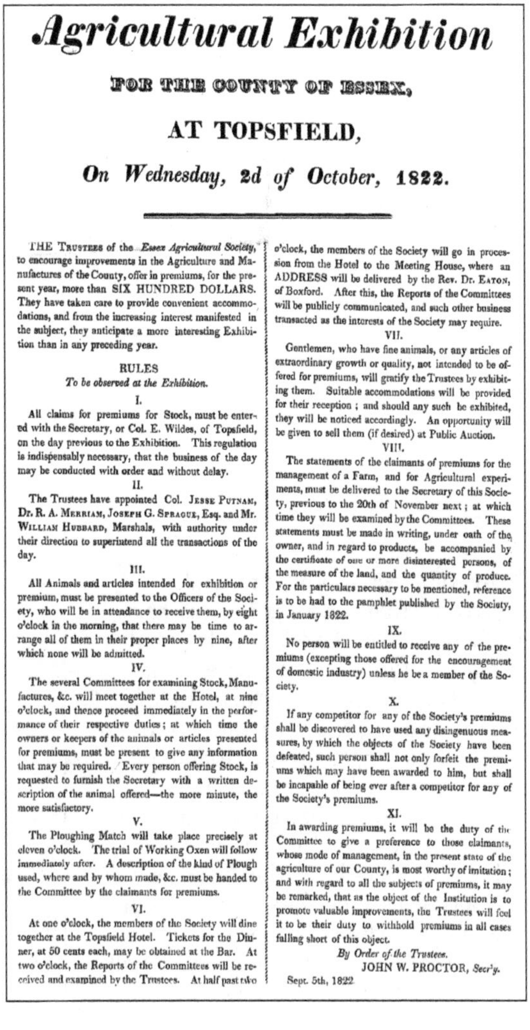

Agricultural Exhibition

FOR THE COUNTY OF ESSEX,

AT TOPSFIELD,

On Wednesday, 2d of October, 1822.

THE TRUSTEES of the *Essex Agricultural Society*, to encourage improvements in the Agriculture and Manufactures of the County, offer in premiums, for the present year, more than SIX HUNDRED DOLLARS. They have taken care to provide convenient accommodations, and from the increasing interest manifested in the subject, they anticipate a more interesting Exhibition than in any preceding year.

RULES

To be observed at the Exhibition.

I.

All claims for premiums for Stock, must be entered with the Secretary, or Col. E. Wildes, of Topsfield, on the day previous to the Exhibition. This regulation is indispensably necessary, that the business of the day may be conducted with order and without delay.

II.

The Trustees have appointed Col. JESSE PUTNAM, Dr. R. A. MERRIAM, JOSEPH G. SPRAGUE, Esq. and Mr. WILLIAM HUBBARD, Marshals, with authority under their direction to superintend all the transactions of the day.

III.

All Animals and articles intended for exhibition or premium, must be presented to the Officers of the Society, who will be in attendance to receive them, by eight o'clock in the morning, that there may be time to arrange all of them in their proper places by nine, after which none will be admitted.

IV.

The several Committees for examining Stock, Manufactures, &c. will meet together at the Hotel, at nine o'clock, and thence proceed immediately in the performance of their respective duties; at which time the owners or keepers of the animals or articles presented for premiums, must be present to give any information that may be required. Every person offering Stock, is requested to furnish the Secretary with a written description of the animal offered—the more minute, the more satisfactory.

V.

The Ploughing Match will take place precisely at eleven o'clock. The trial of Working Oxen will follow immediately after. A description of the kind of Plough used, where and by whom made, &c. must be handed to the Committee by the claimants for premiums.

VI.

At one o'clock, the members of the Society will dine together at the Topsfield Hotel. Tickets for the Dinner, at 50 cents each, may be obtained at the Bar. At two o'clock, the Reports of the Committees will be received and examined by the Trustees. At half past two o'clock, the members of the Society will go in procession from the Hotel to the Meeting House, where an ADDRESS will be delivered by the Rev. Dr. EATON, of Boxford. After this, the Reports of the Committees will be publicly communicated, and such other business transacted as the interests of the Society may require.

VII.

Gentlemen, who have fine animals, or any articles of extraordinary growth or quality, not intended to be offered for premiums, will gratify the Trustees by exhibiting them. Suitable accommodations will be provided for their reception; and should any such be exhibited, they will be noticed accordingly. An opportunity will be given to sell them (if desired) at Public Auction.

VIII.

The statements of the claimants of premiums for the management of a Farm, and for Agricultural experiments, must be delivered to the Secretary of this Society, previous to the 20th of November next; at which time they will be examined by the Committees. These statements must be made in writing, under oath of the owner, and in regard to products, be accompanied by the certificate of one or more disinterested persons, of the measure of the land, and the quantity of produce. For the particulars necessary to be mentioned, reference is to be had to the pamphlet published by the Society, in January 1822.

IX.

No person will be entitled to receive any of the premiums (excepting those offered for the encouragement of domestic industry) unless he be a member of the Society.

X.

If any competitor for any of the Society's premiums shall be discovered to have used any disingenuous measures, by which the objects of the Society have been defeated, such person shall not only forfeit the premiums which may have been awarded to him, but shall be incapable of being ever after a competitor for any of the Society's premiums.

XI.

In awarding premiums, it will be the duty of the Committee to give a preference to those claimants, whose mode of management, in the present state of the agriculture of our County, is most worthy of imitation; and with regard to all the subjects of premiums, it may be remarked, that as the object of the Institution is to promote valuable improvements, the Trustees will feel it to be their duty to withhold premiums in all cases falling short of this object.

By Order of the Trustees.

JOHN W. PROCTOR, *Secr'y.*

Sept. 5th, 1822.

This September 22, 1822 announcement by the trustees of the Essex Agricultural Society informs all interested persons that the society will hold its third annual Agricultural Exhibition at Topsfield on Wednesday, October 2, 1822. The announcement also spells out the rules for the exhibitors. The trustees planned to give out more than $600 in premiums.

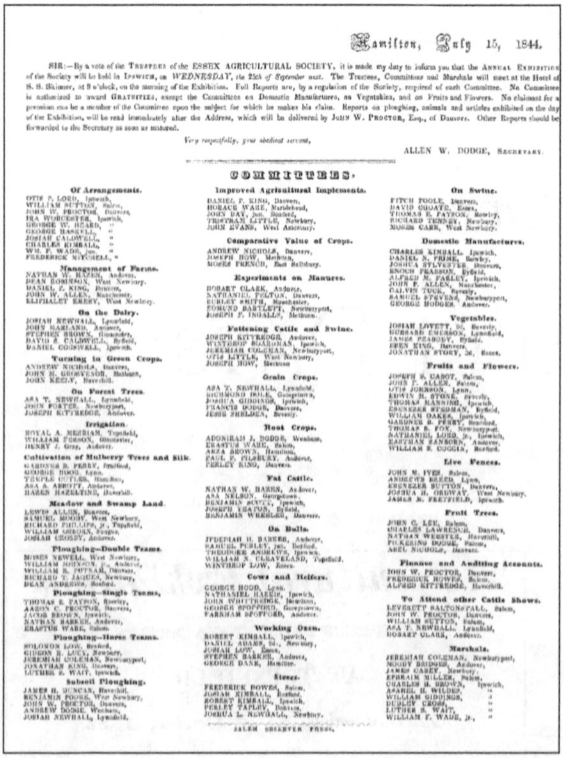

Hamilton, July 15, 1844.

SIR:—By a vote of the TRUSTEES of the ESSEX AGRICULTURAL SOCIETY, it is made my duty to inform you that the ANNUAL EXHIBITION of the Society will be held in IPSWICH, on *WEDNESDAY, the 25th of September next.* The Trustees, Committees and Marshals will meet at the Hotel of S. S. Skinner, at 9 o'clock, on the morning of the Exhibition. Full Reports are, by a regulation of the Society, required of each Committee. No Committee is authorized to award GRATUITIES, except the Committees on Domestic Manufactures, on Vegetables, and on Fruits and Flowers. No claimant for a premium can be a member of the Committee upon the subject for which he makes his claim. Reports on ploughing, animals and articles exhibited on the day of the Exhibition, will be read immediately after the Address, which will be delivered by JOHN W. PROCTOR, Esq., of Danvers. Other Reports should be forwarded to the Secretary as soon as matured.

Very respectfully, your obedient servant,

ALLEN W. DODGE, SECRETARY.

COMMITTEES.

Of Arrangements.
OTIS P. LORD, Ipswich,
WILLIAM SUTTON, Salem,
JOHN W. PROCTOR, Danvers,
IRA WORCESTER, Ipswich,
GEORGE W. HEARD, "
GEORGE HASKELL, "
JOSIAH CALDWELL, "
CHARLES KIMBALL, "
WM. F. WADE, jun. "
FREDERICK MITCHELL, "

Management of Farms.
NATHAN W. HAZEN, Andover,
DEAN ROBINSON, West Newbury,
DANIEL P. KING, Danvers,
JOHN W. ALLEN, Manchester,
ELIPHALET EMERY, West Newbury.

On the Dairy.
JOSIAH NEWHALL, Lynnfield,
JOHN MARLAND, Andover,
STEPHEN BROWN, Gloucester,
DAVID S. CALDWELL, Byfield,
DANIEL COGSWELL, Ipswich.

Turning in Green Crops.
ANDREW NICHOLS, Danvers,
JOHN H. GROSVENOR, Methuen,
JOHN KEELY, Haverhill.

On Forest Trees.
ASA T. NEWHALL, Lynnfield,
JOHN PORTER, Newburyport,
JOSEPH KITTREDGE, Andover.

Irrigation.
ROYAL A. MERRIAM, Topsfield,
WILLIAM PERSON, Gloucester,
HENRY J. Gray, Andover.

Cultivation of Mulberry Trees and Silk.
GARDNER B. PERRY, Bradford,
GEORGE HOOD, Lynn,
TEMPLE CUTLER, Hamilton,
ASA A. ABBOTT, Andover,
HAZEN HAZELTINE, Haverhill.

Meadow and Swamp Land.
LEWIS ALLEN, Danvers,
SAMUEL MOODY, West Newbury,
RICHARD PHILLIPS, jr., Topsfield,
WILLIAM OSBORN, Saugus,
JOSIAH CROSBY, Andover.

Ploughing—Double Teams.
MOSES NEWELL, West Newbury,
WILLIAM JOHNSON, jr., Andover,
WILLIAM R. PUTNAM, Danvers,
RICHARD T. JAQUES, Newbury,
DEAN ANDREWS, Boxford.

Ploughing—Single Teams.
THOMAS E. PAYSON, Rowley,
AARON C. PROCTOR, Danvers,
JACOB BROWN, Ipswich,
NATHAN BARKER, Andover,
ERASTUS WARE, Salem.

Ploughing—Horse Teams.
SOLOMON LOW, Boxford,
GIDEON R. LUCY, Newbury,
JEREMIAH COLEMAN, Newburyport,
JONATHAN KING, Danvers,
LUTHER S. WAIT, Ipswich.

Subsoil Ploughing.
JAMES H. DUNCAN, Haverhill,
BENJAMIN POORE, West Newbury,
JOHN W. PROCTOR, Danvers,
ANDREW DODGE, Wenham,
JOSIAH NEWHALL, Lynnfield.

Improved Agricultural Implements.
DANIEL P. KING, Danvers,
HORACE WARE, Marblehead,
JOHN DAY, jun. Boxford,
TRISTRAM LITTLE, Newbury,
JOHN EVANS, West Amesbury.

Comparative Value of Crops.
ANDREW NICHOLS, Danvers,
JOSEPH HOW, Methuen,
MOSES FRENCH, East Salisbury.

Experiments on Manures.
HOBART CLARK, Andover,
NATHANIEL FELTON, Danvers,
BURLEY SMITH, Manchester,
EDMUND BARTLETT, Newburyport,
JOSEPH F. INGALLS, Methuen.

Fattening Cattle and Swine.
JOSEPH KITTREDGE, Andover,
WINTHROP BOARDMAN, Ipswich,
JEREMIAH COLEMAN, Newburyport,
OTIS LITTLE, West Newbury,
JOSEPH HOW, Methuen.

Grain Crops.
ASA T. NEWHALL, Lynnfield,
RICHMOND DOLE, Georgetown,
JOSHUA GIDDINGS, Ipswich,
FRANCIS DODGE, Danvers,
JESSE SHELDEN, Beverly.

Root Crops.
ADONIRAM J. DODGE, Wenham,
ERASTUS WARE, Salem,
ARZA BROWN, Hamilton,
PAUL P. PILSBURY, Andover,
PERLEY KING, Danvers.

Fat Cattle.
NATHAN W. HAZEN, Andover,
ASA NELSON, Georgetown,
BENJAMIN SCOTT, Ipswich,
JOSEPH YEATON, Byfield,
BENJAMIN WHEELER, Danvers.

On Bulls.
JEDEDIAH H. BARKER, Andover,
SAMUEL PERLEY, jun. Boxford,
THEODORE ANDREWS, Ipswich,
WILLIAM N. CLEAVELAND, Topsfield,
WINTHROP LOW, Essex.

Cows and Heifers.
GEORGE HOOD, Lynn,
NATHANIEL HARRIS, Ipswich,
JOHN WHITTREDGE, Hamilton,
GEORGE SPOFFORD, Georgetown,
FARNHAM SPOFFORD, Andover.

Working Oxen.
ROBERT KIMBALL, Ipswich,
DANIEL ADAMS, 3d., Newbury,
JOSIAH LOW, Essex,
STEPHEN BARKER, Andover,
GEORGE DANE, Hamilton.

Steers.
FREDERICK HOWES, Salem,
JOSIAH KIMBALL, Boxford,
ROBERT KIMBALL, Ipswich,
PERLEY TAPLEY, Danvers,
JOSHUA L. NEWHALL, Newbury.

On Swine.
FITCH POOLE, Danvers,
DAVID CHOATE, Essex,
THOMAS E. PAYSON, Rowley,
RICHARD TENNEY, Newbury,
MOSES CARR, West Newbury.

Domestic Manufactures.
CHARLES KIMBALL, Ipswich,
DANIEL N. PRIME, Rowley,
JOSHUA SYLVESTER, Danvers,
ENOCH PEARSON, Byfield,
ALFRED M. FARLEY, Ipswich,
JOHN P. ALLEN, Manchester,
CALVIN TUCK, Beverly,
SAMUEL STEVENS, Newburyport,
GEORGE HODGES, Andover.

Vegetables.
JOSIAH LOVETT, 2d, Beverly,
HUBBARD EMERSON, Lynnfield,
JAMES PEABODY, Byfield,
EBEN KING, Danvers,
JONATHAN STORY, 3d, Essex.

Fruits and Flowers.
JOSEPH S. CABOT, Salem,
JOHN P. ALLEN, Salem,
OTIS JOHNSON, Lynn,
EDWIN M. STONE, Beverly,
THOMAS MANNING, Ipswich,
EBENEZER STEDMAN, Byfield,
WILLIAM OAKES, Ipswich,
GARDNER B. PERRY, Bradford,
THOMAS B. FOX, Newburyport,
NATHANIEL LORD, jr., Ipswich,
EASTMAN SANBORN, Andover,
WILLIAM S. COGGIN, Boxford.

Live Fences.
JOHN M. IVES, Salem,
ANDREWS BREED, Lynn,
EBENEZER SUTTON, Danvers,
JOSHUA H. ORDWAY, West Newbury,
JAMES N. FEETFIELD, Ipswich.

Fruit Trees.
JOHN C. LEE, Salem,
CHARLES LAWRENCE, Danvers,
NATHAN WEBSTER, Haverhill,
PICKERING DODGE, Salem,
ABEL NICHOLS, Danvers.

Finance and Auditing Accounts.
JOHN W. PROCTOR, Danvers,
FREDERICK HOWES, Salem,
ALFRED KITTREDGE, Haverhill.

To Attend other Cattle Shows.
LEVERETT SALTONSTALL, Salem,
JOHN W. PROCTOR, Danvers,
WILLIAM SUTTON, Salem,
ASA T. NEWHALL, Lynnfield,
HOBART CLARK, Andover.

Marshals.
JEREMIAH COLEMAN, Newburyport,
MOODY BRIDGES, Andover,
JAMES CAREY, Newbury,
EPHRAIM MILLER, Salem,
CHARLES H. BROWN, Ipswich,
ASAHEL H. WILDES, "
WILLIAM GIDDINGS, "
DUDLEY CROSS, "
LUTHER S. WAIT, "
WILLIAM F. WADE, jr., "

SALEM OBSERVER PRESS.

On July 15, 1844, the trustees published this announcement indicating to all readers that the exhibition would be held in Andover on September 25. It also listed all the committees and the members of each of these committees.

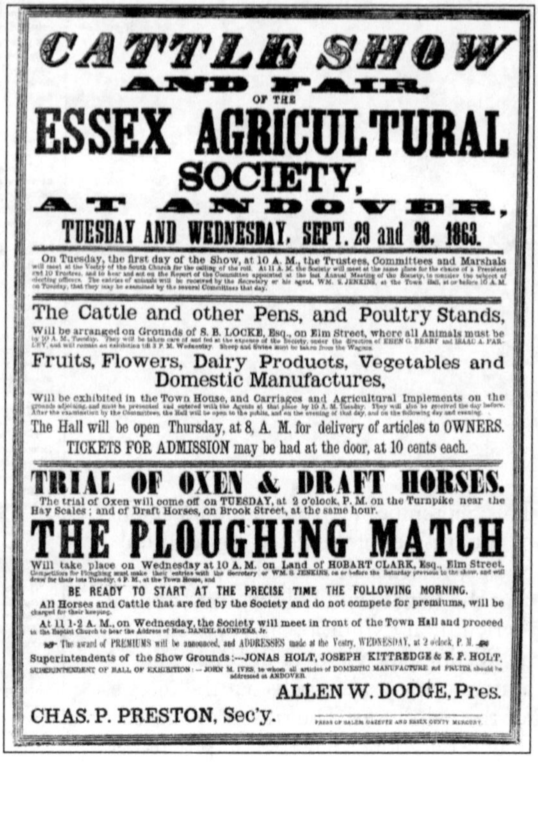

CATTLE SHOW

AND FAIR

OF THE

ESSEX AGRICULTURAL SOCIETY,

AT ANDOVER,

TUESDAY AND WEDNESDAY, SEPT. 29 and 30, 1863.

On Tuesday, the first day of the Show, at 10 A. M., the Trustees, Committees and Marshals will meet at the Vestry of the South Church for the calling of the roll. At 11 A. M. the Society will meet at the same place for the choice of a President and 10 Trustees, and to hear and act on the Report of the Committee appointed at the last Annual Meeting of the Society, to consider the subject of electing officers. The entries of animals will be received by the Secretary or his agent, WM. S. JENKINS, at the Town Hall, at or before 10 A. M. on Tuesday, that they may be examined by the several Committees that day.

The Cattle and other Pens, and Poultry Stands,

Will be arranged on Grounds of S. B. LOCKE, Esq., on Elm Street, where all Animals must be by 10 A. M., Tuesday. They will be taken care of and fed at the expense of the Society, under the direction of EBEN G. BERRY and ISAAC A. FARLEY, and will remain on exhibition till 3 P. M. Wednesday. Sheep and Swine must be taken from the Wagons.

Fruits, Flowers, Dairy Products, Vegetables and Domestic Manufactures,

Will be exhibited in the Town House, and Carriages and Agricultural Implements on the grounds adjoining, and must be presented and entered with the Agents at that place by 10 A. M. Tuesday. They will also be received the day before. After the examination by the Committees, the Hall will be open to the public, and on the evening of that day, and on the following day and evening.

The Hall will be open Thursday, at 8, A. M. for delivery of articles to OWNERS.

TICKETS FOR ADMISSION may be had at the door, at 10 cents each.

TRIAL OF OXEN & DRAFT HORSES.

The trial of Oxen will come off on TUESDAY, at 2 o'clock, P. M. on the Turnpike near the Hay Scales; and of Draft Horses, on Brook Street, at the same hour.

THE PLOUGHING MATCH

Will take place on Wednesday at 10 A. M. on Land of HOBART CLARK, Esq., Elm Street. Competitors for Ploughing must make their entries with the Secretary or WM. S. JENKINS, on or before the Saturday previous to the show, and will draw for their lots Tuesday, 4 P. M., at the Town House, and

BE READY TO START AT THE PRECISE TIME THE FOLLOWING MORNING.

All Horses and Cattle that are fed by the Society and do not compete for premiums, will be charged for their keeping.

At 11 1-2 A. M., on Wednesday, the Society will meet in front of the Town Hall and proceed to the Baptist Church to hear the Address of Hon. DANIEL SAUNDERS, Jr.

The award of PREMIUMS will be announced, and ADDRESSES made at the Vestry, WEDNESDAY, at 2 o'clock, P. M.

Superintendents of the Show Grounds:—JONAS HOLT, JOSEPH KITTREDGE & E. F. HOLT.

SUPERINTENDENT OF HALL OF EXHIBITION:—JOHN M. IVES, to whom all articles of DOMESTIC MANUFACTURE and FRUITS should be addressed at ANDOVER.

ALLEN W. DODGE, Pres.

CHAS. P. PRESTON, Sec'y.

PRESS OF SALEM GAZETTE AND ESSEX COUNTY MERCURY.

Pictured here is another broadside advertising that the Essex County Fair would be held in Andover on September 29 and 30, 1863. The smaller exhibits were to be located in the "town hall," and the animal pens and poultry stands were to be arranged on the grounds of S.B. Locke, Esq., on Elm Street. The trial of oxen was to be on the "turnpike" near the "hay scales" at 2:00 on Tuesday. Other broadsides listed the departments and outlined all the classes and rules of each for the exhibitors. The society has several of these two- by three-and-a-half-foot broadsides in its archives.

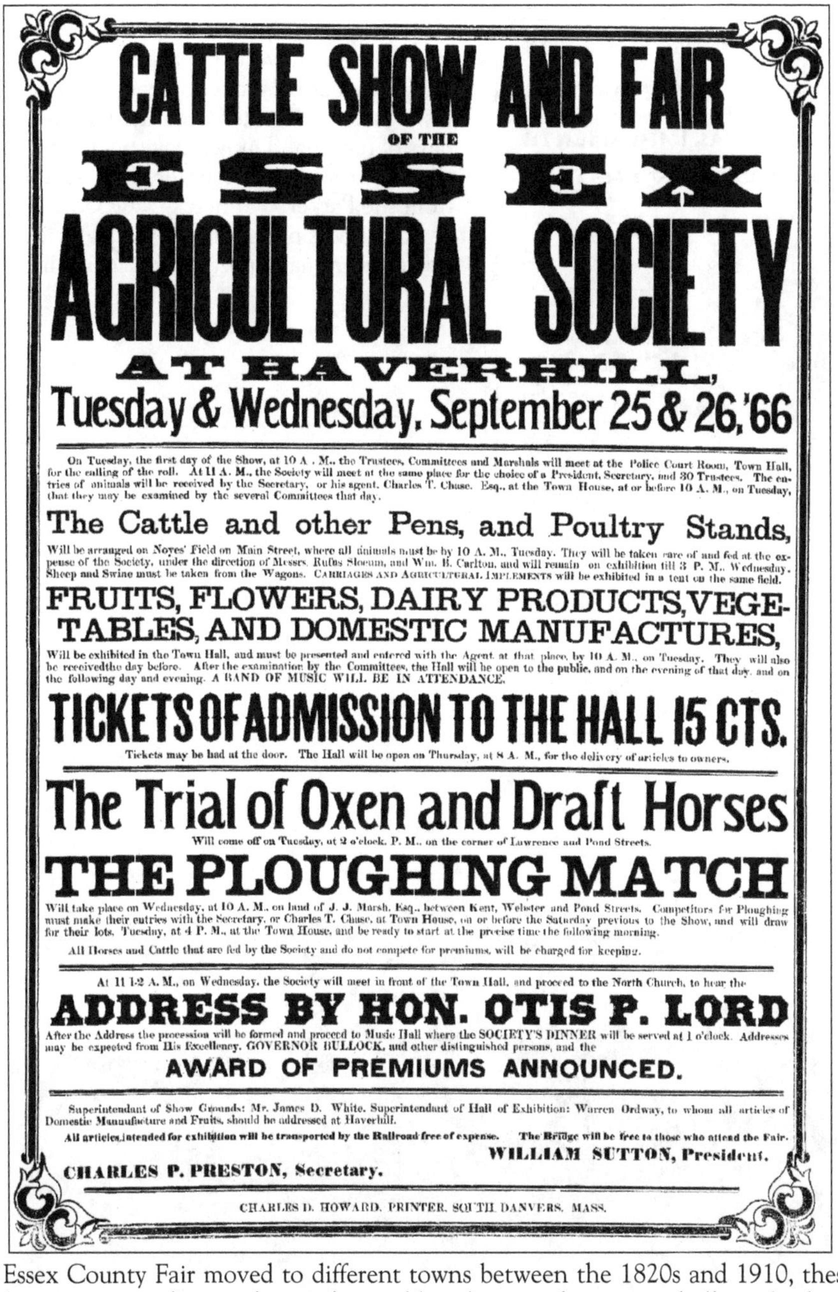

CATTLE SHOW AND FAIR

OF THE

ESSEX

AGRICULTURAL SOCIETY

AT HAVERHILL,

Tuesday & Wednesday, September 25 & 26,'66

On Tuesday, the first day of the Show, at 10 A. M., the Trustees, Committees and Marshals will meet at the Police Court Room, Town Hall, for the calling of the roll. At 11 A. M., the Society will meet at the same place for the choice of a President, Secretary, and 30 Trustees. The entries of animals will be received by the Secretary, or his agent, Charles T. Chase, Esq., at the Town House, at or before 10 A. M., on Tuesday, that they may be examined by the several Committees that day.

The Cattle and other Pens, and Poultry Stands,

Will be arranged on Noyes' Field on Main Street, where all animals must be by 10 A. M., Tuesday. They will be taken care of and fed at the expense of the Society, under the direction of Messrs. Rufus Slocum, and Wm. B. Carlton, and will remain on exhibition till 3 P. M., Wednesday. Sheep and Swine must be taken from the Wagons. CARRIAGES AND AGRICULTURAL IMPLEMENTS will be exhibited in a tent on the same field.

FRUITS, FLOWERS, DAIRY PRODUCTS, VEGETABLES, AND DOMESTIC MANUFACTURES,

Will be exhibited in the Town Hall, and must be presented and entered with the Agent at that place, by 10 A. M., on Tuesday. They will also be receivedthe day before. After the examination by the Committees, the Hall will be open to the public, and on the evening of that day, and on the following day and evening. A BAND OF MUSIC WILL BE IN ATTENDANCE.

TICKETS OF ADMISSION TO THE HALL 15 CTS.

Tickets may be had at the door. The Hall will be open on Thursday, at 8 A. M., for the delivery of articles to owners.

The Trial of Oxen and Draft Horses

Will come off on Tuesday, at 2 o'clock, P. M., on the corner of Lawrence and Pond Streets.

THE PLOUGHING MATCH

Will take place on Wednesday, at 10 A. M., on land of J. J. Marsh, Esq., between Kent, Webster and Pond Streets. Competitors for Ploughing must make their entries with the Secretary, or Charles T. Chase, at Town House, on or before the Saturday previous to the Show, and will draw for their lots, Tuesday, at 4 P. M., at the Town House, and be ready to start at the precise time the following morning.

All Horses and Cattle that are fed by the Society and do not compete for premiums, will be charged for keeping.

At 11 1-2 A. M., on Wednesday, the Society will meet in front of the Town Hall, and proceed to the North Church, to hear the

ADDRESS BY HON. OTIS P. LORD

After the Address the procession will be formed and proceed to Music Hall where the SOCIETY'S DINNER will be served at 1 o'clock. Addresses may be expected from His Excellency, GOVERNOR BULLOCK, and other distinguished persons, and the

AWARD OF PREMIUMS ANNOUNCED.

Superintendant of Show Grounds: Mr. James D. White. Superintendant of Hall of Exhibition: Warren Ordway, to whom all articles of Domestic Manufacture and Fruits, should be addressed at Haverhill.

All articles intended for exhibition will be transported by the Railroad free of expense. The Bridge will be free to those who attend the Fair.

WILLIAM SUTTON, President.

CHARLES P. PRESTON, Secretary.

CHARLES D. HOWARD, PRINTER, SOUTH DANVERS, MASS.

As the Essex County Fair moved to different towns between the 1820s and 1910, these large broadsides were printed up and posted in public places such as town halls and other public buildings, railway stations, stores, billboards, and anywhere the public could view them. The broadside above announces that the "Cattle Show and Fair" was to be held on September 26 and 26, 1866, at Haverhill. One of the highlights at all of these early agricultural fairs was the oxen and horse pulling contests. The "ploughing matches" were also a big part of the competitions. The flowers, fruits, vegetables, and domestic manufacture were usually exhibited in the town halls, while the animal pens and poultry stands were located either on the common or on a nearby farmer's field. At this exhibition, the pens were arranged on Noyes' Field on Main Street.

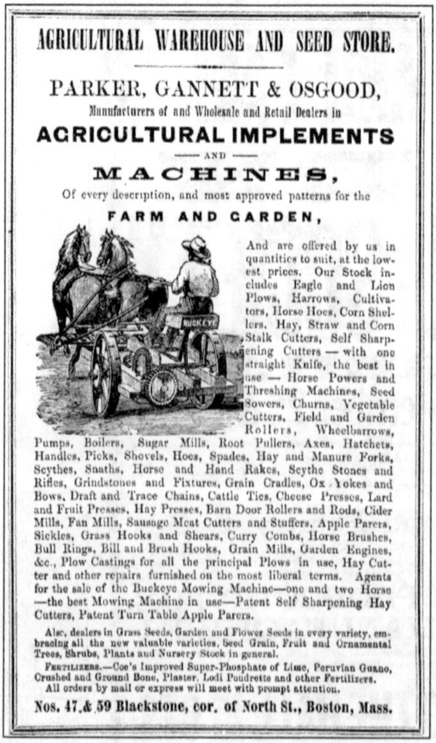

AGRICULTURAL WAREHOUSE AND SEED STORE.

PARKER, GANNETT & OSGOOD,

Manufacturers of and Wholesale and Retail Dealers in

AGRICULTURAL IMPLEMENTS

— AND —

MACHINES,

Of every description, and most approved patterns for the

FARM AND GARDEN,

And are offered by us in quantities to suit, at the lowest prices. Our Stock includes Eagle and Lion Plows, Harrows, Cultivators, Horse Hoes, Corn Shellers, Hay, Straw and Corn Stalk Cutters, Self Sharpening Cutters — with one straight Knife, the best in use — Horse Powers and Threshing Machines, Seed Sowers, Churns, Vegetable Cutters, Field and Garden Rollers, Wheelbarrows, Pumps, Boilers, Sugar Mills, Root Pullers, Axes, Hatchets, Handles, Picks, Shovels, Hoes, Spades, Hay and Manure Forks, Scythes, Snaths, Horse and Hand Rakes, Scythe Stones and Rifles, Grindstones and Fixtures, Grain Cradles, Ox Yokes and Bows, Draft and Trace Chains, Cattle Ties, Cheese Presses, Lard and Fruit Presses, Hay Presses, Barn Door Rollers and Rods, Cider Mills, Fan Mills, Sausage Meat Cutters and Stuffers, Apple Parers, Sickles, Grass Hooks and Shears, Curry Combs, Horse Brushes, Bull Rings, Bill and Brush Hooks, Grain Mills, Garden Engines, &c., Plow Castings for all the principal Plows in use, Hay Cutter and other repairs furnished on the most liberal terms. Agents for the sale of the Buckeye Mowing Machine—one and two Horse—the best Mowing Machine in use—Patent Self Sharpening Hay Cutters, Patent Turn Table Apple Parers.

Also, dealers in Grass Seeds, Garden and Flower Seeds in every variety, embracing all the new valuable varieties, Seed Grain, Fruit and Ornamental Trees, Shrubs, Plants and Nursery Stock in general.

Fertilizers.—Coe's Improved Super-Phosphate of Lime, Peruvian Guano, Crushed and Ground Bone, Plaster, Lodi Poudrette and other Fertilizers.

All orders by mail or express will meet with prompt attention.

Nos. 47 & 59 Blackstone, cor. of North St., Boston, Mass.

Agricultural equipment dealers around New England were advertising the new horse-drawn mowing machines that had been developed in the last decade or so. This freed the farm help from days of hard hand mowing with a scythe to get the hay crops in much faster. This advertisement from the Parker, Gannett & Osgood Company of Boston was placed in the Essex Agricultural Society's annual report, *Transactions*, in 1862.

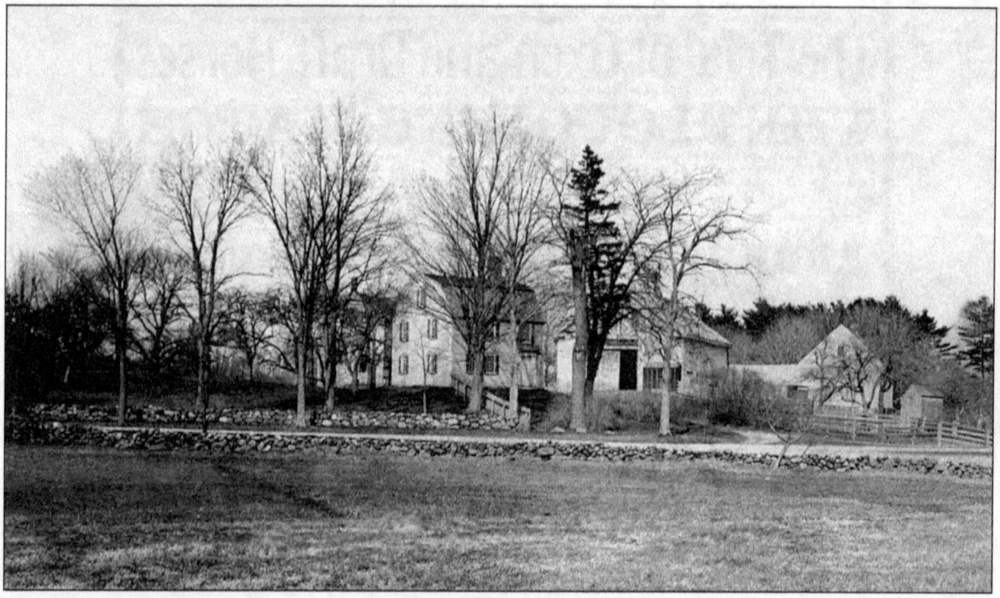

The Treadwell Farm became the present site of the fair in 1910, but the Essex Agricultural Society had owned it since 1856. The Dexter-Treadwell house was originally built by a Dr. Dexter from Danvers in 1741. A new barn was contracted to be built by the society in 1859 on the society's experimental farm on South Main Street in Topsfield. This great photograph was sent in 1906 to a former tenant on the property in the 1890s by her cousin George Stanwood of Topsfield. The site of the buildings is now the Essex County Cooperative, and the original Dexter house was moved in 1974 to Ipswich Road by Johnathan Bond. (Photograph courtesy of Johnathan Bond.)

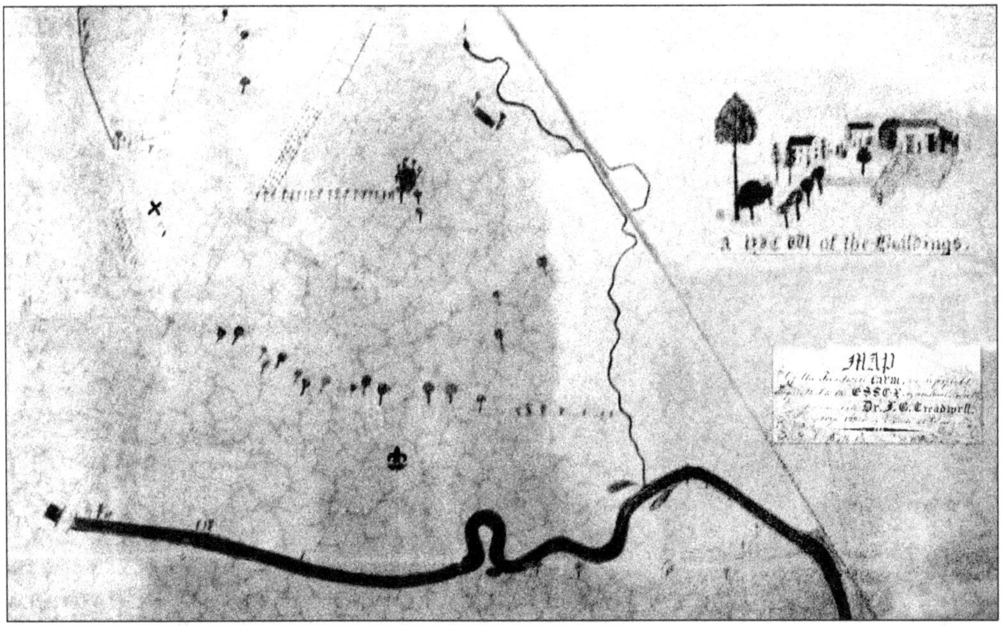

Drawn in the late 1800s, this map of the Treadwell Farm shows the location of the Dexter-Treadwell house. Note the inset of the house and buildings. The X shows the location of these farm buildings and the Dexter house. The Treadwell Farm became the site of the fair in 1910, although the society had either controlled or owned it since 1856, when it was bequeathed to the society. At one time, cranberries were grown on this farm, as it had many wet boggy areas, which were required for growing cranberries. (Map courtesy of the Peabody-Essex Institute.)

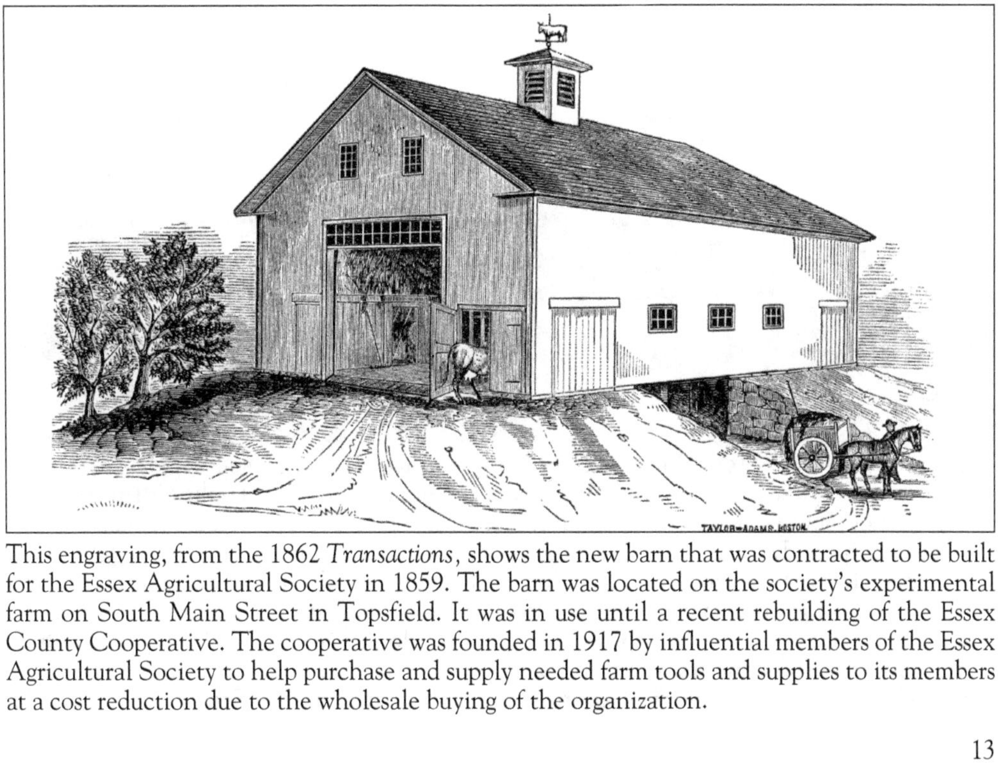

This engraving, from the 1862 *Transactions*, shows the new barn that was contracted to be built for the Essex Agricultural Society in 1859. The barn was located on the society's experimental farm on South Main Street in Topsfield. It was in use until a recent rebuilding of the Essex County Cooperative. The cooperative was founded in 1917 by influential members of the Essex Agricultural Society to help purchase and supply needed farm tools and supplies to its members at a cost reduction due to the wholesale buying of the organization.

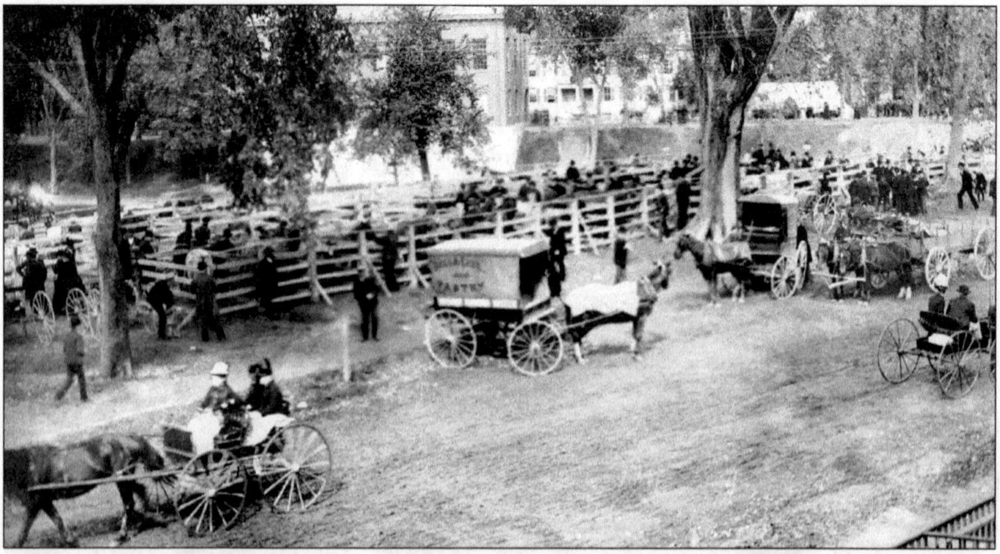

Before 1895, the Essex County Fair was moved from town to town, usually stopping in each for two consecutive years before going to a new town in the county. In 1885 and 1886, the fair was held in Newburyport on "the Mall" behind the courthouse, which can be seen here in the background. Most of the animal pens seem to contain horses that are on exhibition. In the very distant background, a tent is set up on the far side of High Street, most likely housing the poultry stands. (Photograph courtesy of the Newburyport Archival Center at the Newburyport Public Library.)

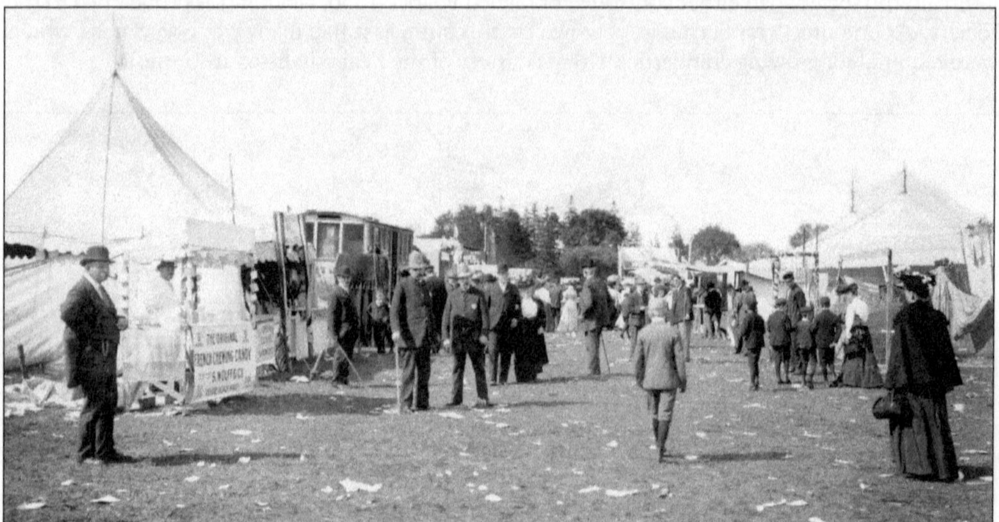

The fairs of the Essex Agricultural Society from 1895 to 1909 were held in Peabody on the society's 50-acre show grounds. In this 1903 view, we see "Fakers Row." To the far left is a stand selling "the original French Chewing Candy" for 5¢ a bag or 10¢ a box. The manufacturer of this chewy taffy candy was S. Wolf & Company of Revere Beach. Other stands are selling cigars, sandwiches, cider, Moxie, and other cold drinks. According to advertisements in later programs for the fair, the Moxie Company also bottled Pureoxia, a pale dry or golden ginger ale. The "night lunch wagon" on the left appears to be a White House Cafe model. The wagon, with its "Wild West" painted vignettes, was possibly a product of the H.T. Buckley Car Manufacturing Company of Worcester. (Photograph courtesy of the Peabody Historical Society.)

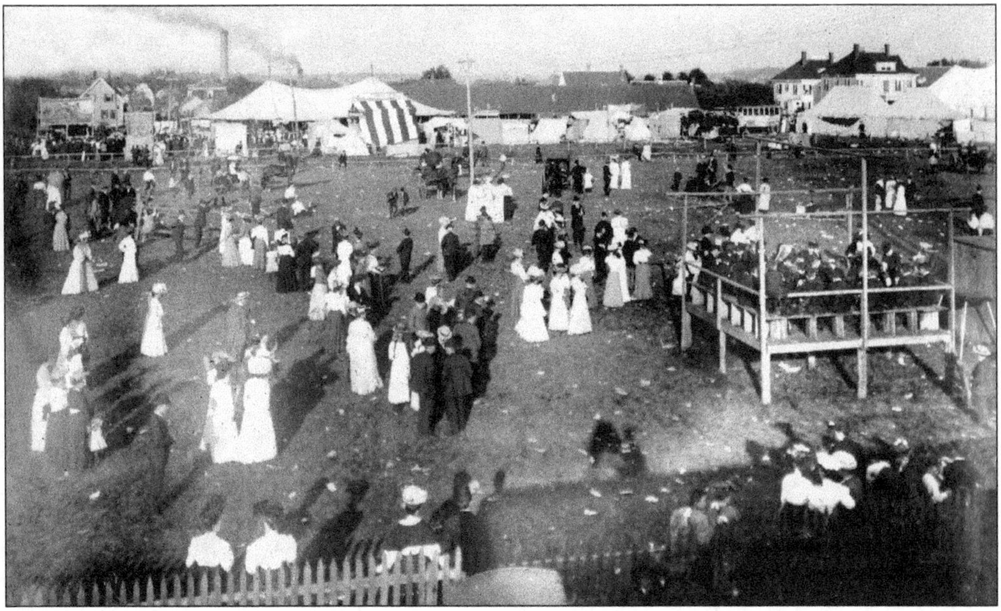

The horse show at the Essex County Fair, featured in this 1903 photograph taken in Peabody, has a carriage class in progress while the crowds of ladies and gentlemen walk about them in their finest Sunday dress. Fairs were not held on Sundays, however, until the 1950s. In the depths of this photograph can be seen a high-wheeled bicycle leaning against a "night lunch wagon" on Fakers Row. Note the banner with likenesses of pugilists on it; the fairs also featured boxing matches. The tall smokestack in the background is that of the A.C. Lawrence Leather Company. (Photograph courtesy of the Peabody Historical Society.)

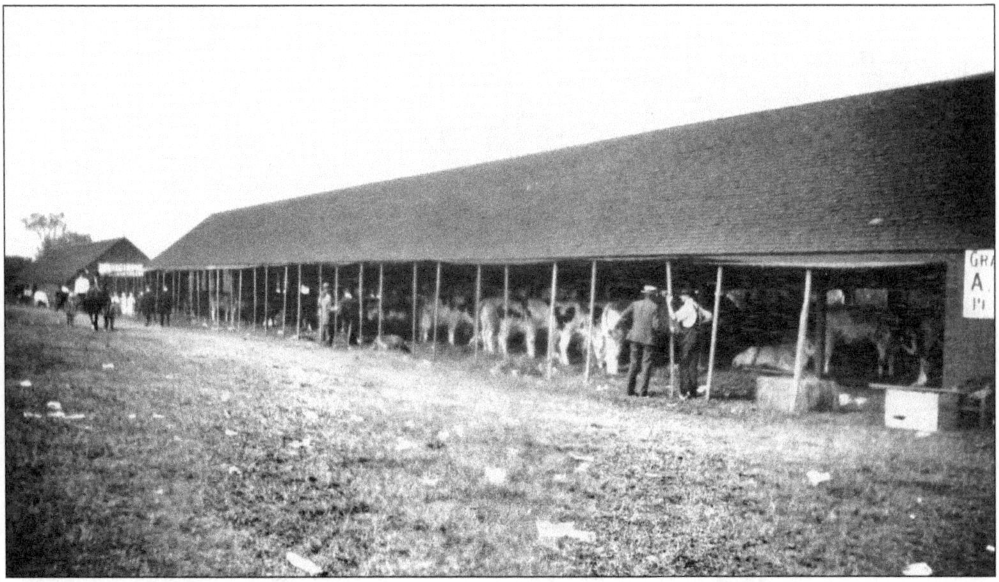

In 1903, some forward-looking photographer recorded the horse and cattle sheds of the Essex Agricultural Society's fairgrounds in Peabody. The society owned 50 acres upon which it erected several buildings and a racetrack as a home for the fair in Peabody. Note that the large sign on the near end of the shed is advertising grain for sale by A.P. Ames & Company of Peabody. (Photograph courtesy of the Peabody Historical Society.)

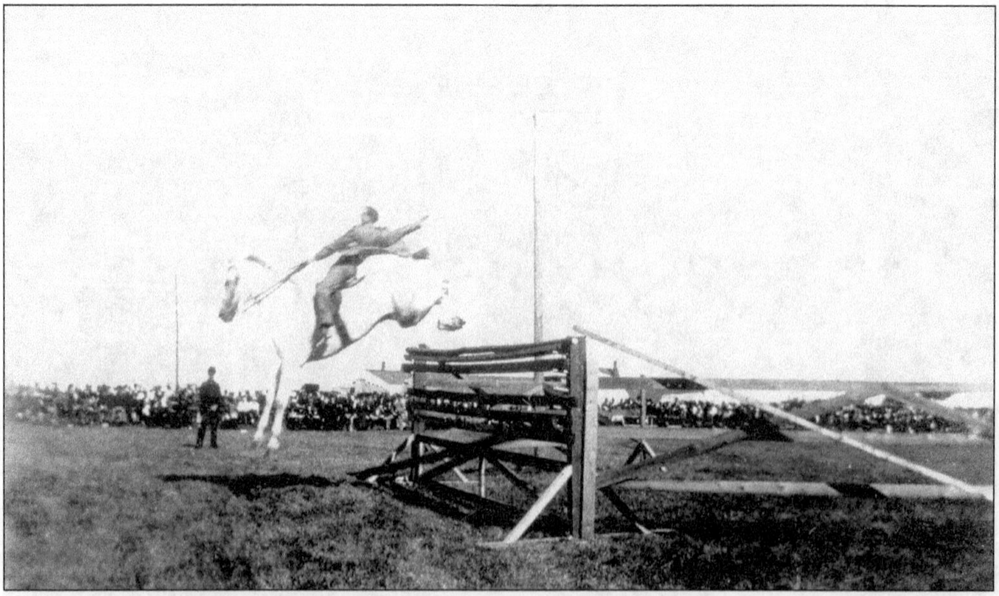

At the fair's horse show in Peabody in 1903, Tornado (owned by T.W. Pierce) won the hurdle-jumping class. The Essex County Fair, as it was called before 1927, was held on the society's property in Peabody from 1985 through 1909. In 1910, the society voted to sell the 50-acre site and move the fair back to Topsfield to its Treadwell Farm location. (Photograph courtesy of the Peabody Historical Society.)

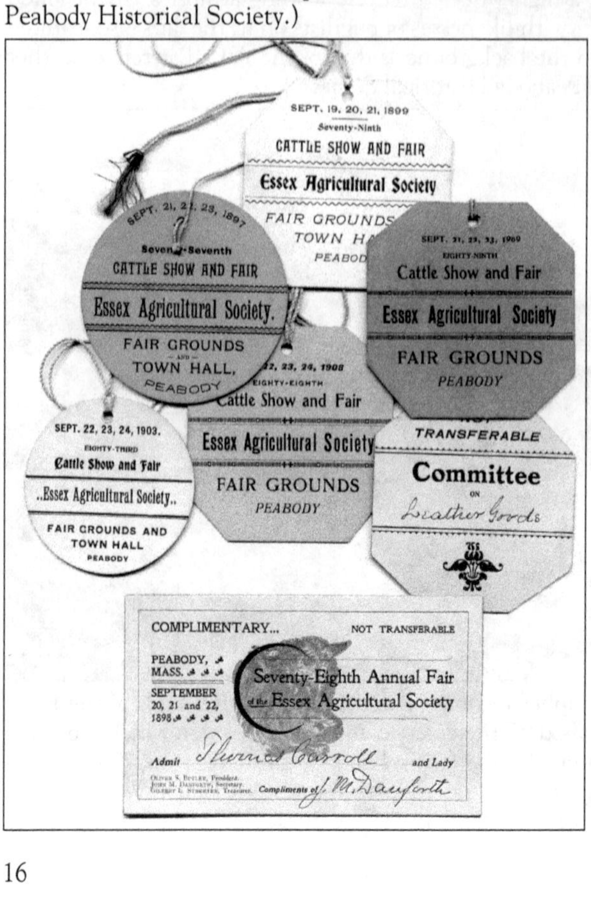

The Peabody Historical Society has several of these interesting old tags and exhibitors and committee passes dating from 1897 to 1909, the last year the fair was held in Peabody. (Photograph courtesy of the Peabody Historical Society.)

When the fair moved permanently to Topsfield in 1910, these smaller admission tags were used. The admission fee was 25¢ instead of the 15¢ charged before 1900. Some of these tags in the society's archives have the word "school" stamped on the back. Presumably, schoolchildren were charged a smaller admission fee, as is the case today.

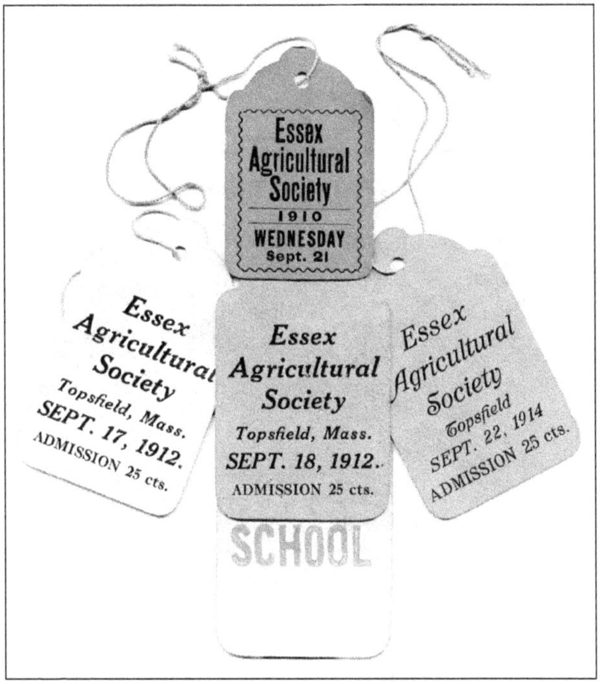

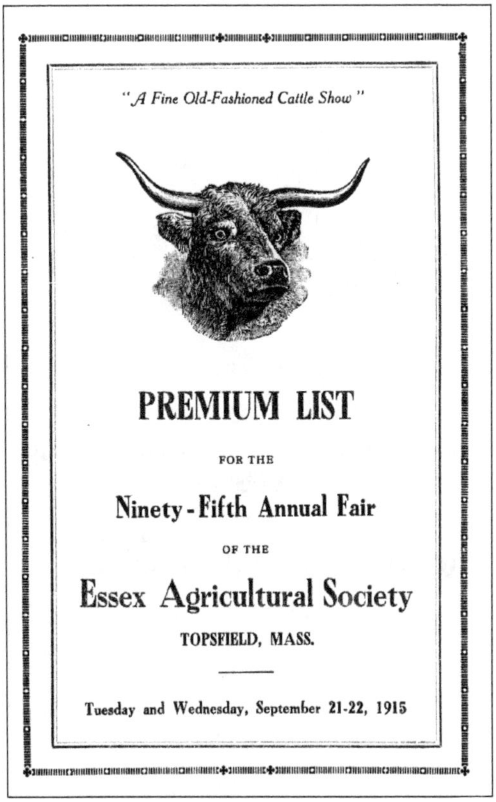

"A Fine Old-Fashioned Cattle Show"

PREMIUM LIST

FOR THE

Ninety-Fifth Annual Fair

OF THE

Essex Agricultural Society

TOPSFIELD, MASS.

Tuesday and Wednesday, September 21-22, 1915

Each year, the fair printed premium books to give exhibitors planning to enter various exhibits in the fair. Some premium books over the years were specific to one department only, such as this 1915 book for the cattle show. Many of these earlier premium books and programs give us a window into the types of activities that were held at the fairs in the past. They sometimes give very detailed information, such as classes, rules, awards available, and so on.

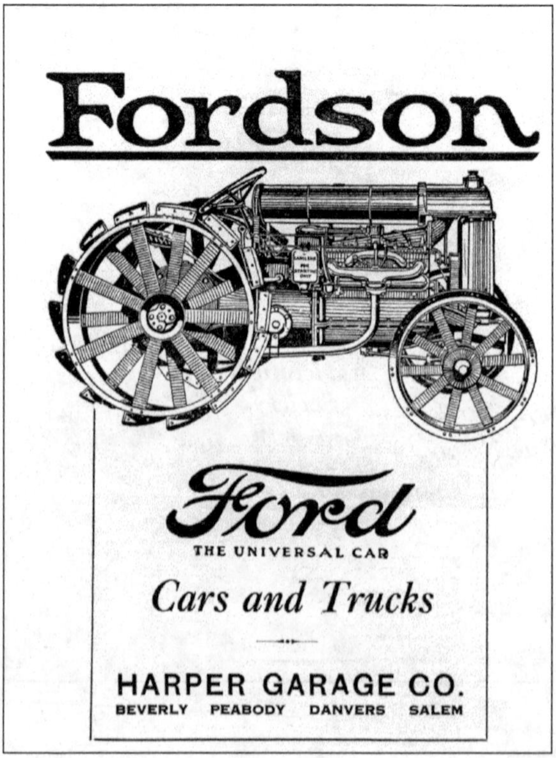

The Harper Garage Company of Beverly, Peabody, Danvers, and Salem sold Ford cars and trucks, as well as Fordson tractors. This simple but effective advertisement appeared in the 1921 Topsfield Fair premium book. A great photograph of the company's Fordson tractor display, with accompanying implements and attachments, can also be viewed in this volume.

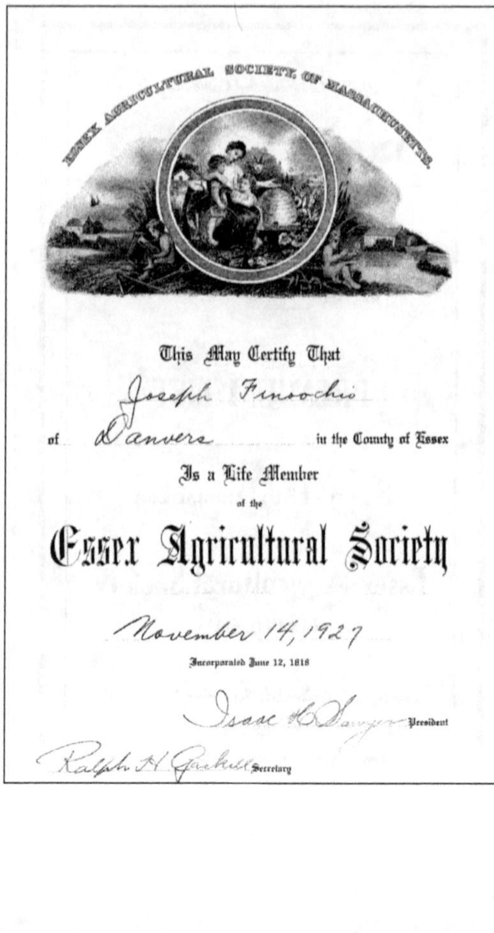

Perhaps the oldest living life member of the Essex Agricultural Society is 96-year-old Joseph Finoochio of Topsfield. Finoochio was born in 1907 in Italy and attended the Essex Agricultural School in Hathorne after moving to America as a child. He became a life member of the society upon his graduation in 1927, as his membership certificate indicates. He still plays tennis about every day and works in his gardens all season long. His passion is growing roses around his home. (Certificate courtesy of Joseph Finoochio.)

Two

THE FAIRGROUNDS

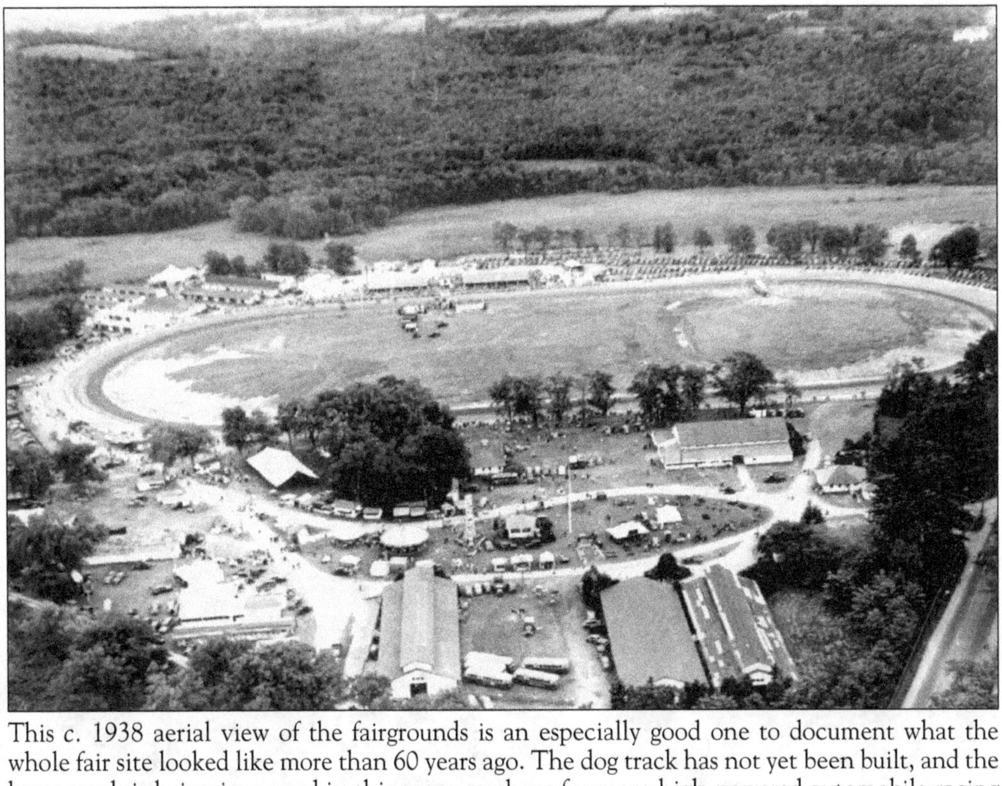

This *c.* 1938 aerial view of the fairgrounds is an especially good one to document what the whole fair site looked like more than 60 years ago. The dog track has not yet been built, and the horse track is being improved in this scene, perhaps for more high-powered automobile racing starting this year. The large horse stable tents have been replaced by wooden buildings on the Ipswich River side of the track, possibly for horses or racecar accommodations. The patches on the plant and flower building roof indicate that the fair's maintenance crew has some more work ahead of them. The 30- by 60-foot Women's Building was built near the Grange building and is now the bee and honey building, which until Coolidge Hall was completed, served as the home arts building. This building can be seen near the center to the right of the grove of trees where once stood the old Topsfield Block House, built to protect residents from Native American raids. The addition to the Women's Building on its southern, or racetrack, side has not yet been constructed.

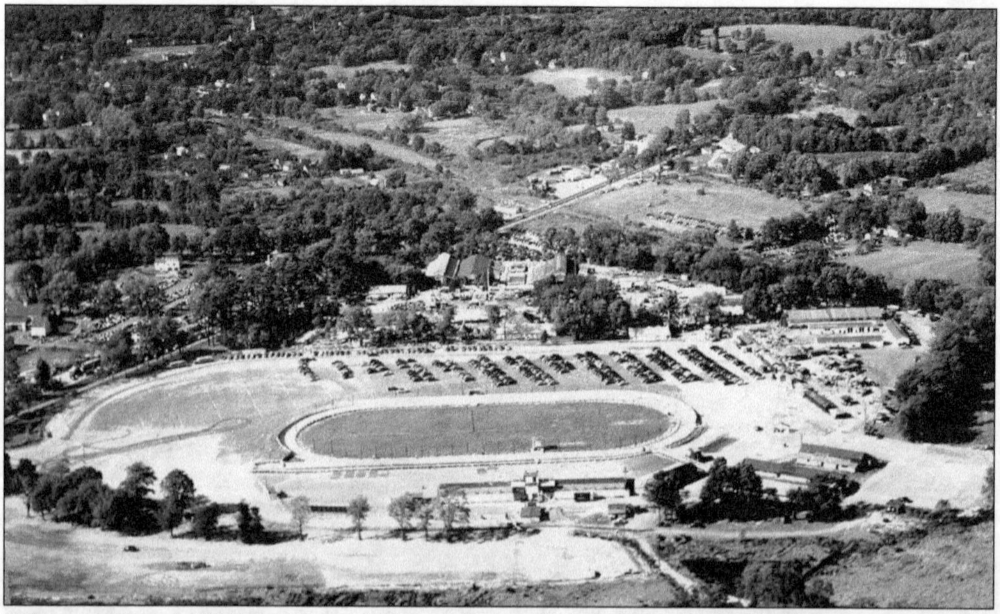

Taken in the early 1950s, this view looking northwest shows the racetrack area in the foreground and a relatively small midway area as compared to today's midway. Apparently, there is to be no horse racing on this day, as the racetrack area and grandstands seem to be deserted. Between the three large exhibition halls in the center are the swimming pool and bleachers used for the Water Follies shows.

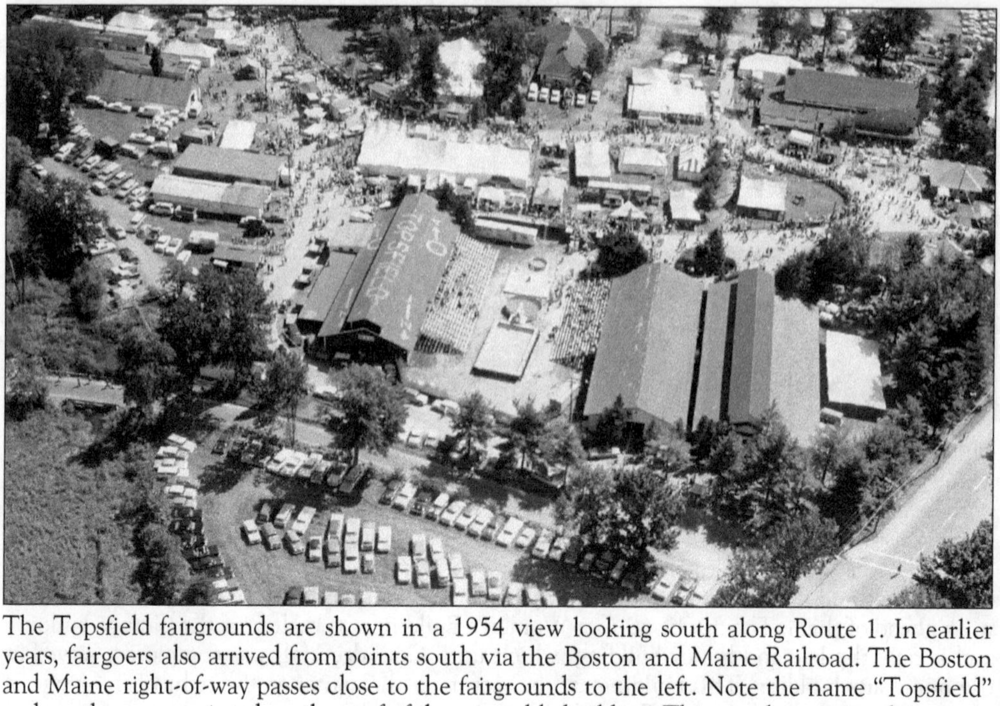

The Topsfield fairgrounds are shown in a 1954 view looking south along Route 1. In earlier years, fairgoers also arrived from points south via the Boston and Maine Railroad. The Boston and Maine right-of-way passes close to the fairgrounds to the left. Note the name "Topsfield" and north arrows painted on the roof of the vegetable building. This was done on roofs in many towns in the 1950s to help pilots ascertain their location. The arrow with the number four presumably means that it is four miles to the Beverly or nearest airport.

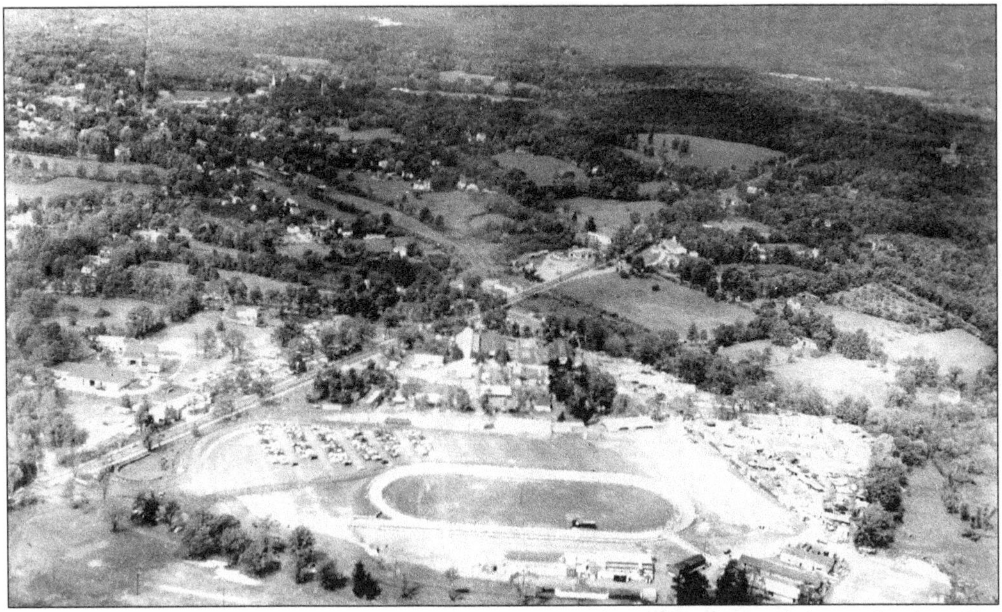

An early-1950s aerial view looking northwest shows a relatively quiet weekday at the fair. The crowds are small, and no races seem to be scheduled.

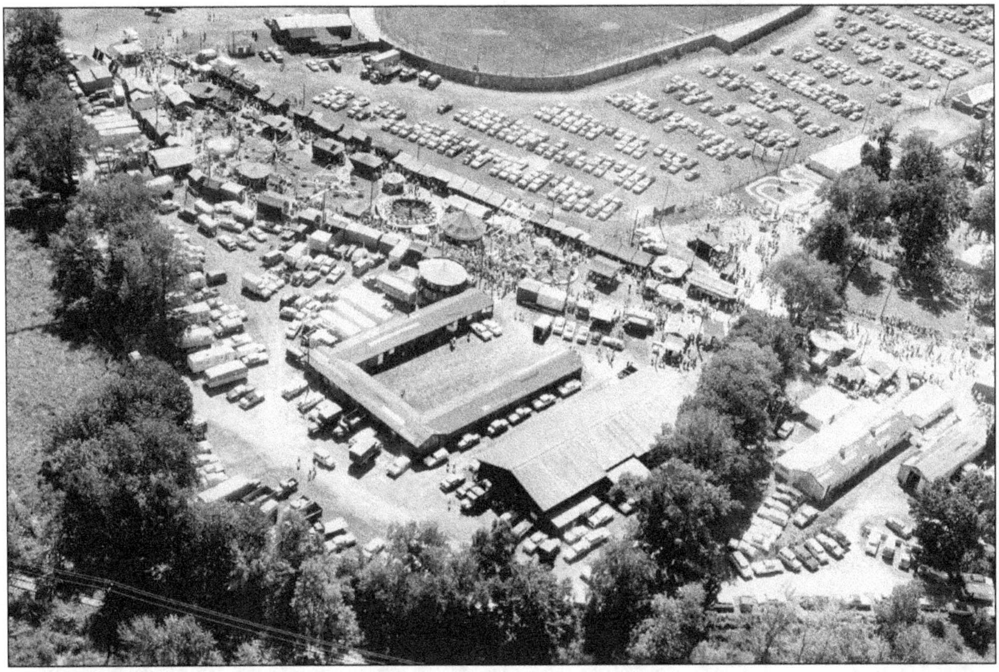

Another fine aerial photograph of the fairgrounds with the dog racing track was taken c. 1962 looking south toward the Ipswich River, bordering the south end of the fair's main parking lot. The U-shaped building in the foreground, built by the Guernsey Breeders Association, was the main cattle barn at that time, with many livestock trucks parked behind it. The new but smaller metal cattle building used today was not constructed until 1964. In more recent years, cattle numbers declined as, one by one, the New England farms went out of business and agricultural production in the whole Northeast grew smaller.

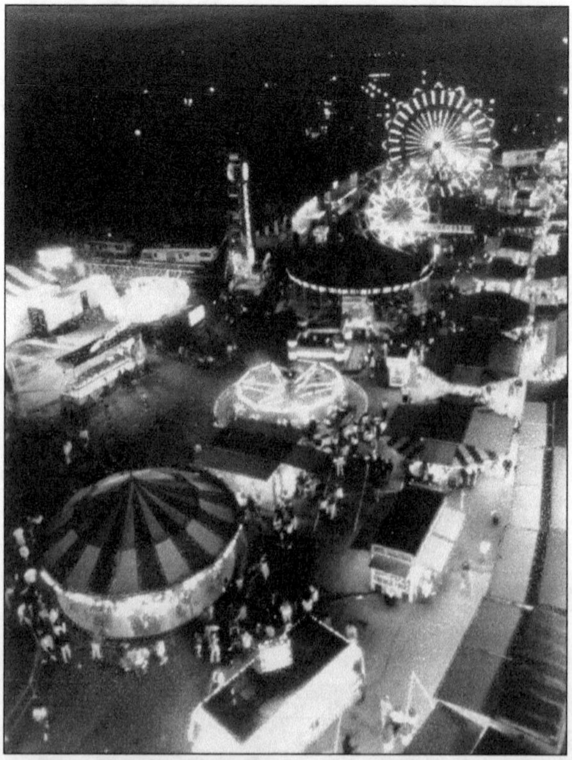

A recent night photograph shows part of the midway where thousands of young people enjoy the fair, eating, playing the games for prizes, and riding on the many wild rides available there. The size and scope of Fakers Row, as the midway used to be called, has changed greatly over the last 100 years.

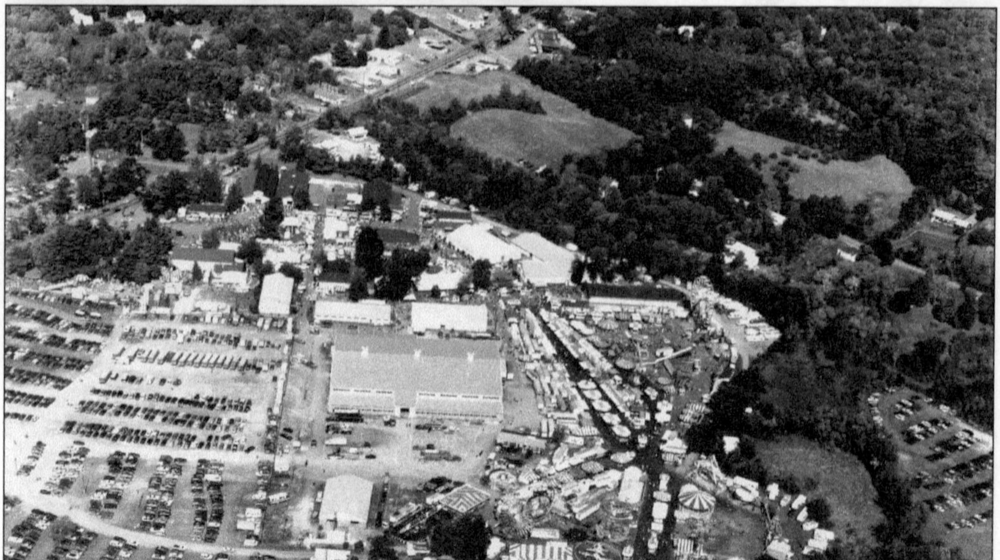

Another great aerial view of the fairgrounds looks north *c.* 1995. Taken decades after the aerial view on the previous page was taken, this photograph shows a new livestock arena and a more extensive midway. Note also that all signs of racetracks have now vanished from the scene except for the little building to the right corner of the arena, which was the old racing clubhouse and racing show secretary's office. This must be a weekday, as the parking lot contains many school buses that have carried loads of schoolchildren to the fair.

Several prominent rides, such as the Super Loop, the Sea Dragon, and the Parachute Ride, are shown in this 1985 aerial photograph of the midway. Note that the Route 97 gate and accompanying parking lot is empty, so it must have been a slow day at the fair. One can see here the old Boston and Maine Railroad roadbed, which was used to bring fair patrons to Topsfield by train in the 1920s and 1930s.

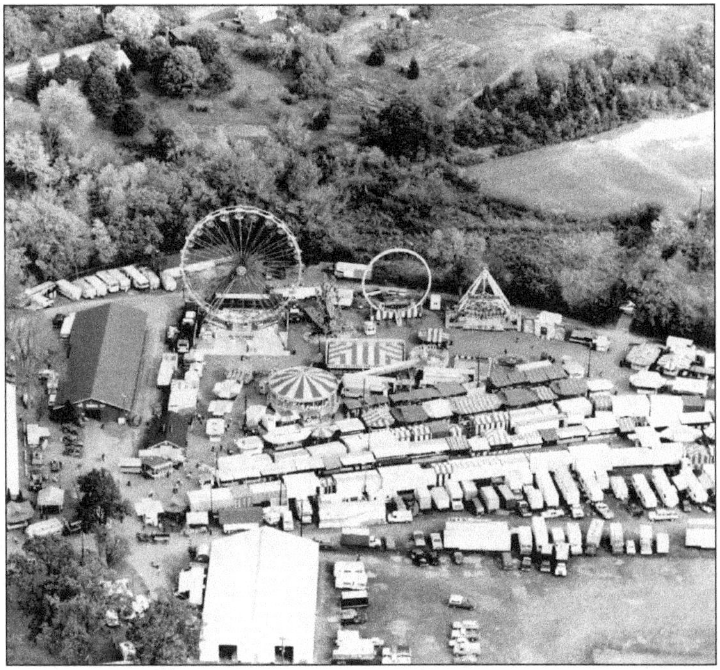

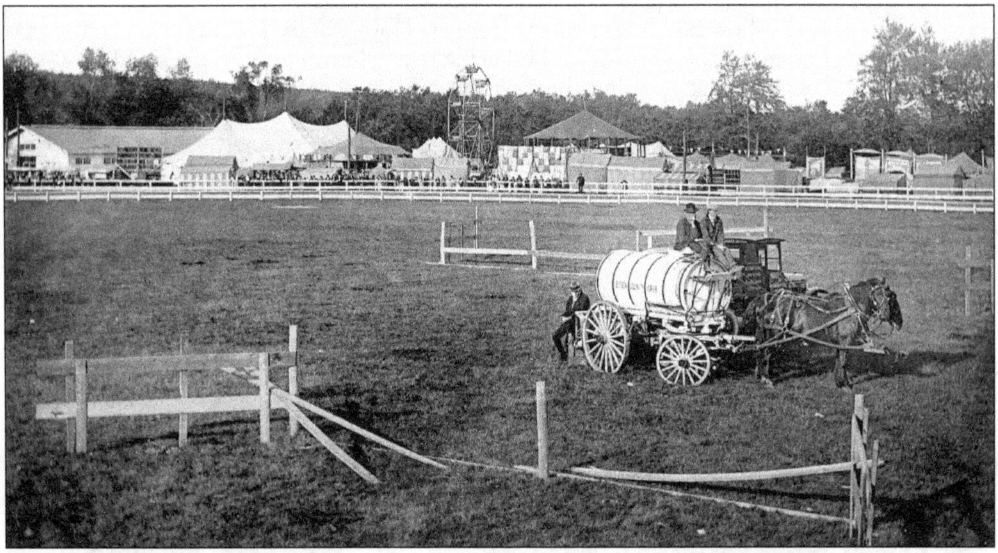

This 1924 photograph shows a large livestock barn behind the racetrack. The dark round canvas roof just to the right of the Ferris wheel covers the barrel-shaped tank of the "Silo-drome," where daring motorcycle riders performed by riding on the vertical walls. The water wagon in the foreground, used daily to wet down the dusty racetrack, was built by Studebaker, and its remains were rescued by Alvin Craig of Essex, the present fair manager, and rebuilt again as a hitch wagon for use in draft horse shows. The heavy truck behind the wagon advertises "McCormick-Deering & International Harvester Power Farm Equipment." The water wagon crew is relaxing in the infield between races apparently, as is the policeman behind them on the track. He is most likely posted there to help keep excited fair patrons from coming onto the racecourse. The name "Essex County Fair" was used for many years by the Essex Agricultural Society for its annual fall fair before today's official name, "Topsfield Fair," was first applied to the event in 1927.

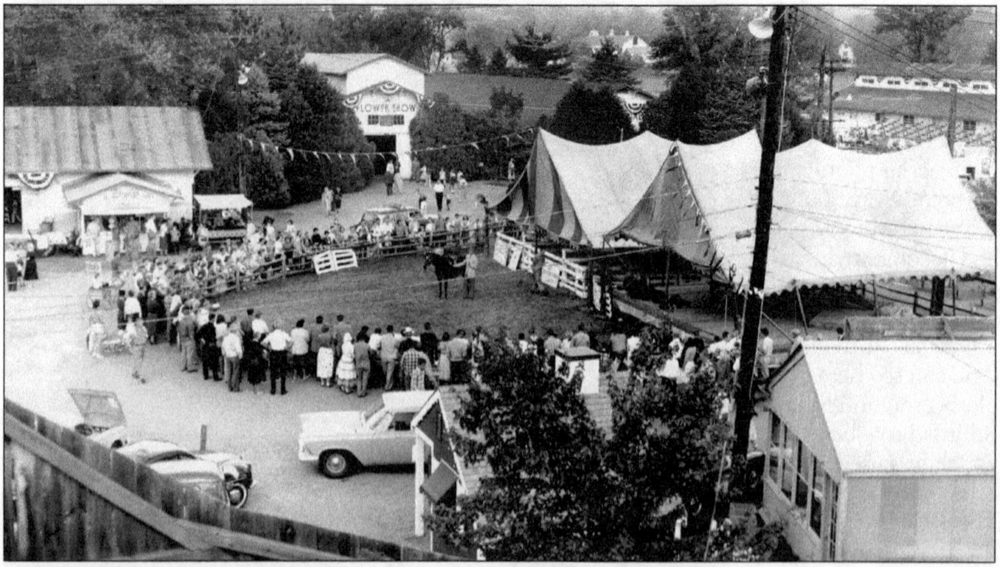

In 1955, Daniel Hurld's third year hosting the public at his Old Essex County Farm (later called the Old Essex County Barnyard), we see him talking about one of his favorite animals, the saddle horse. This is a particularly good view from the Grange building roof. It shows a large crowd gathering around Hurld's farm exhibit. Just to the left of the oval is the 4-H Club building (now the sportsmen's show building) and the Danvers Holy Trinity Methodist Church's food stand, featuring frankforts, lobster rolls, and homemade pie. Just in front of the stand, a woman is driving a goat to a cart advertising the dairy goats. (Photograph courtesy of Barbara and Daniel Hurld.)

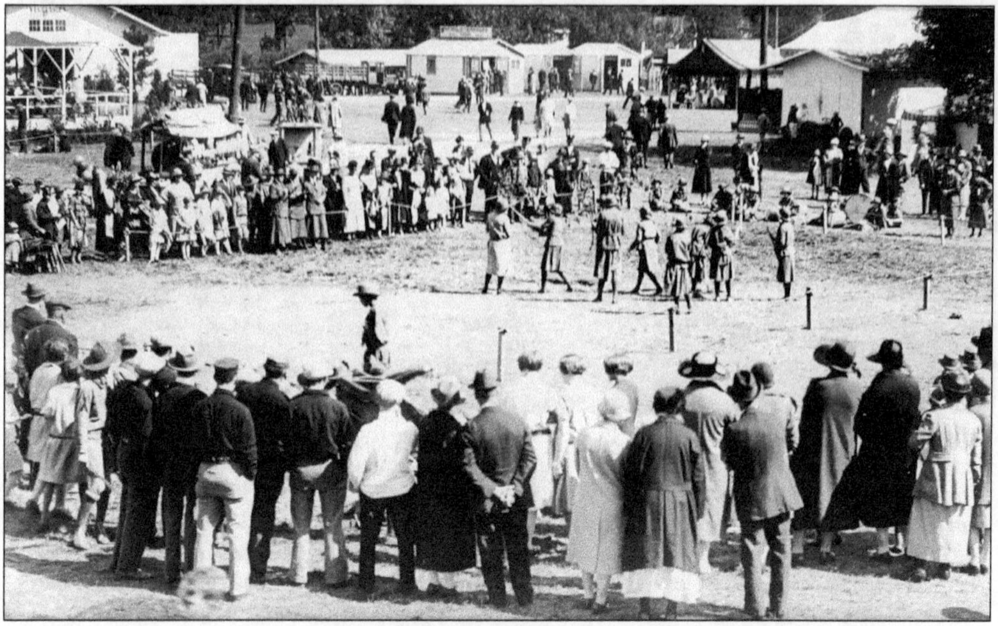

In the center oval in front of the vegetable building we see a group of uniformly dressed women ready to demonstrate some golf shots to a large gathering of spectators on a bright sunny day at the fair in 1926. Note that many of the women are wearing dresses or skirts and the broad-rimmed hats of that period—a contrast to the jeans worn by present-day fairgoers.

Three

The 1920s and 1930s

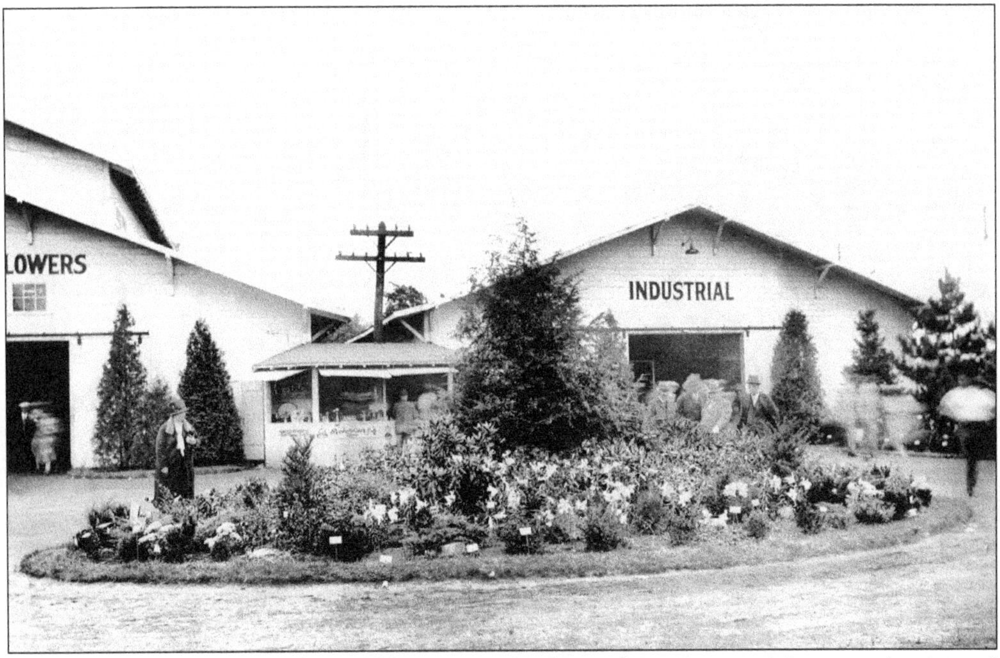

A beautiful planting of shrubs and lilies graces the front of the plant and flower and industrial buildings in 1928 by Cherry Hill Nurseries of West Newbury. The industrial building shown here must have been vacated for that purpose some years after this photograph, as the poultry department moved into it after leaving a similar building toward the back of the fairgrounds to the rabbit department. The refreshment stand behind the garden is advertising Rockefeller cigars and Moxie.

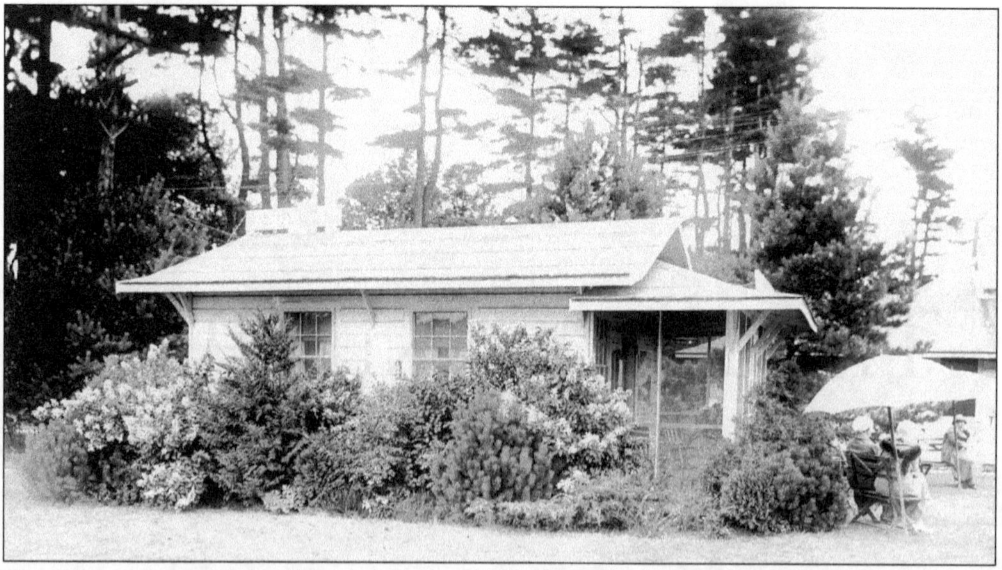

The Christian Science building was constructed just inside the main gate in 1921 along with the dance pavilion, which is just visible next to the Christian Science building in this 1920s photograph. The dance pavilion was also the 4-H building for some time and is now used for the sportsmen's show. There were dances in that pavilion all summer long in the 1920s and 1930s. It was a very popular place for couples to gather on a warm summer evening. Later, as the dance crowds grew, the Women's Building, now the bee building, was used for such social events.

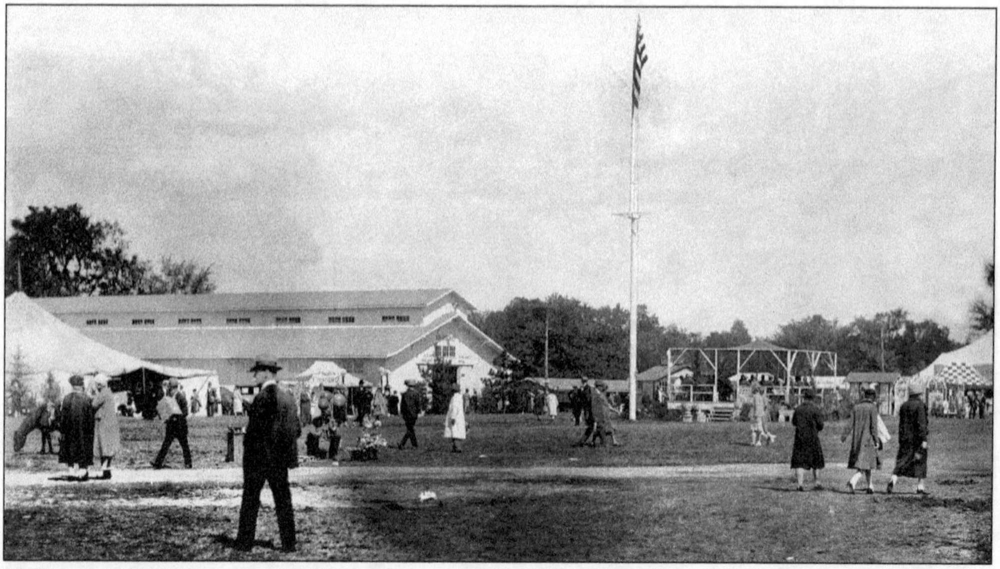

Here we see another view of the main oval of buildings in 1929, showing the center flagpole and band pavilion and the relatively new vegetable building. The vegetable building was constructed in 1923 and, according to the signs on each side of the main doors, housed 4-H, Scout craft, bee, and school exhibits. This building still houses the fruit and vegetable exhibits today. To the left of the large building is the striped tent where gasoline products are on display, and at the back of the oval are several other commercial exhibits in small tents. The large tent to the left may hold some more of the horticultural exhibits.

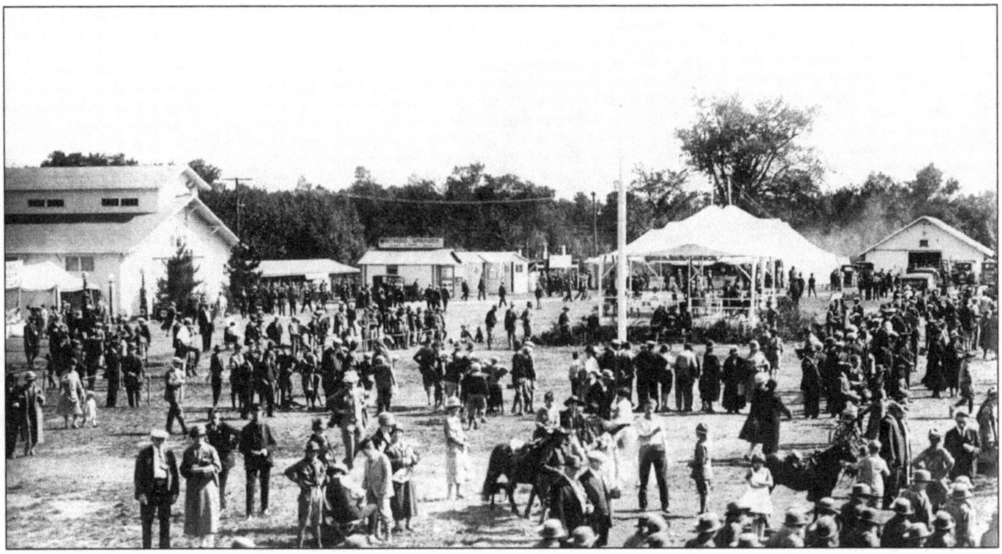

The center oval of the fairgrounds is a very busy place on this day in 1926 with all the people dressed rather fine as if they were going off to church rather than to a fair. This was not, however, a Sunday; the fair was not held on Sundays until after World War II. The large 150- by 50-foot vegetable building (left) was used as a cattle building in the 1950s and, in the 1980s, was again used for fruit and vegetable exhibits.

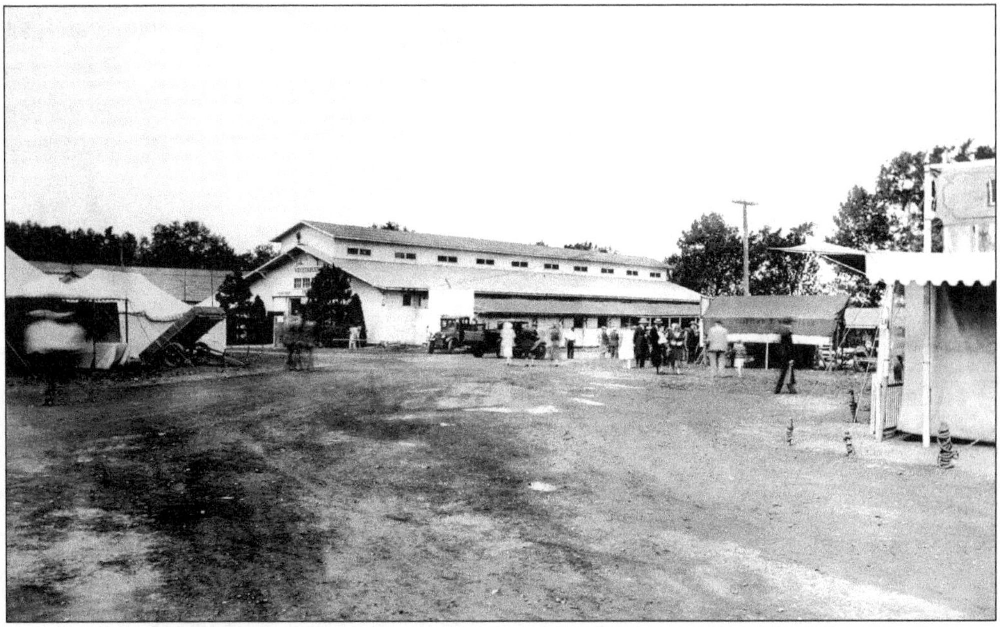

A 1932 view of the vegetable building shows the individual goat exhibit stalls on the east side of the building where the maintenance shops are now located. This is one of the two original 1920s-era fair buildings to have a clerestory-type cupola roof. Clerestory roofs were popular for more than 100 years in railroad cars to help clear smoke and heat from the car interiors as the trains sped along the tracks. The windows on this building did not open, but they did provide extra light. In the 1980s, this upper roof collapsed from a load of snow, and it was not rebuilt in its original style. Only two small ventilation cupolas are on this building today.

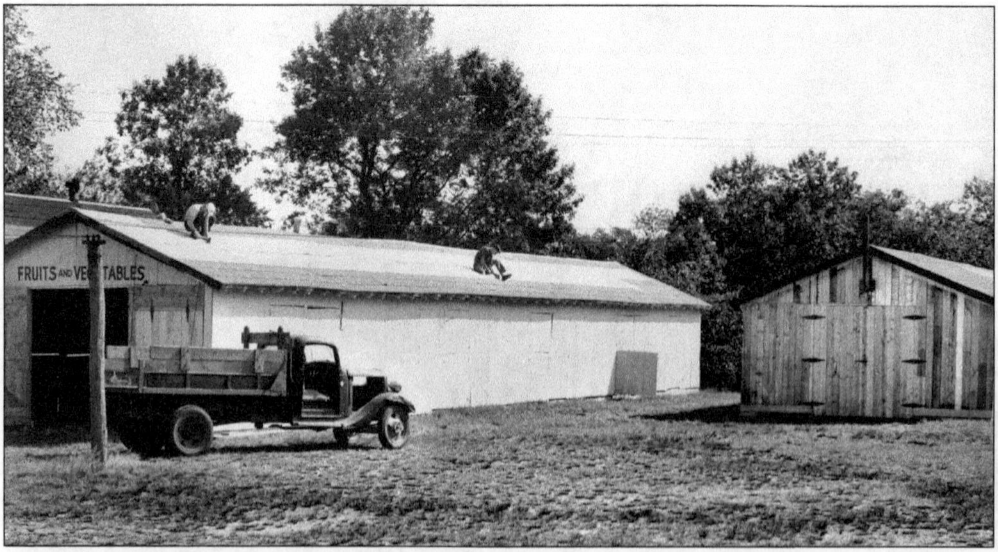

The first vegetable building is being repaired in this photograph taken sometime in the 1930s. The fair maintenance department has chosen to reroof only part of the building to save some money. The structure sometimes served as a cattle building; other photographs reveal a loading dock at the rear for unloading and loading the cattle. Sometime in late 1945 or early 1946, the building burned, as remembered by Harold Kelly of Danvers. It was cleaned up by Kelly and others from Brooksby Farms, where he worked after serving in the war. He later became an instructor and department head in pomology (fruit culture) at the Essex Agricultural School in Hathorne.

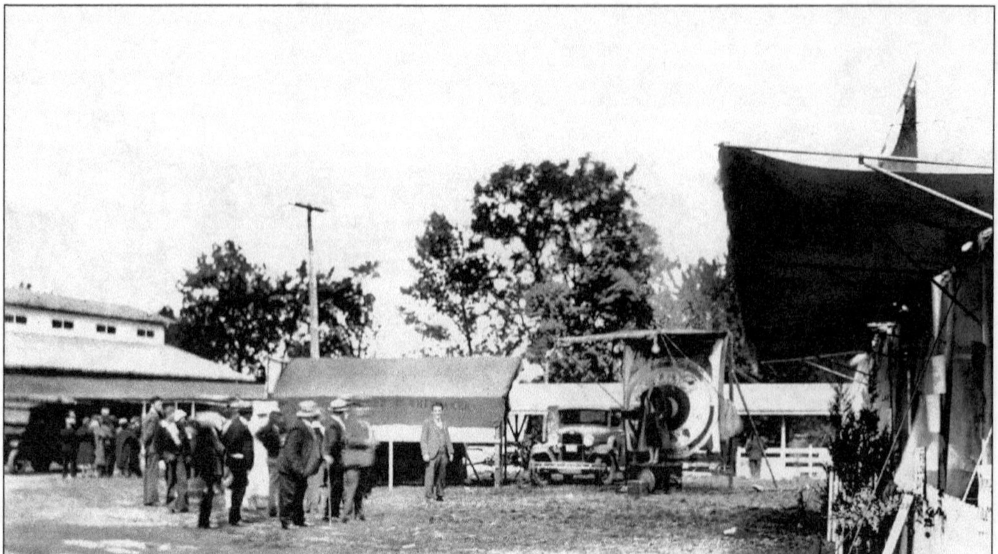

In 1932, the fair was well on its way to being the longest-operating agricultural fair in America. Many types of booths are operated at each fair. In this photograph, a woman operates some sort of a fortune-telling booth. To the right are several types of food booths, including those serving lamb, beef, pork, and chicken dishes. Much of the food at old-time fairs was homemade and a delight to eat. Some home-cooked foods can still be found at fairs today, particularly at some of the church booths and the 4-H building.

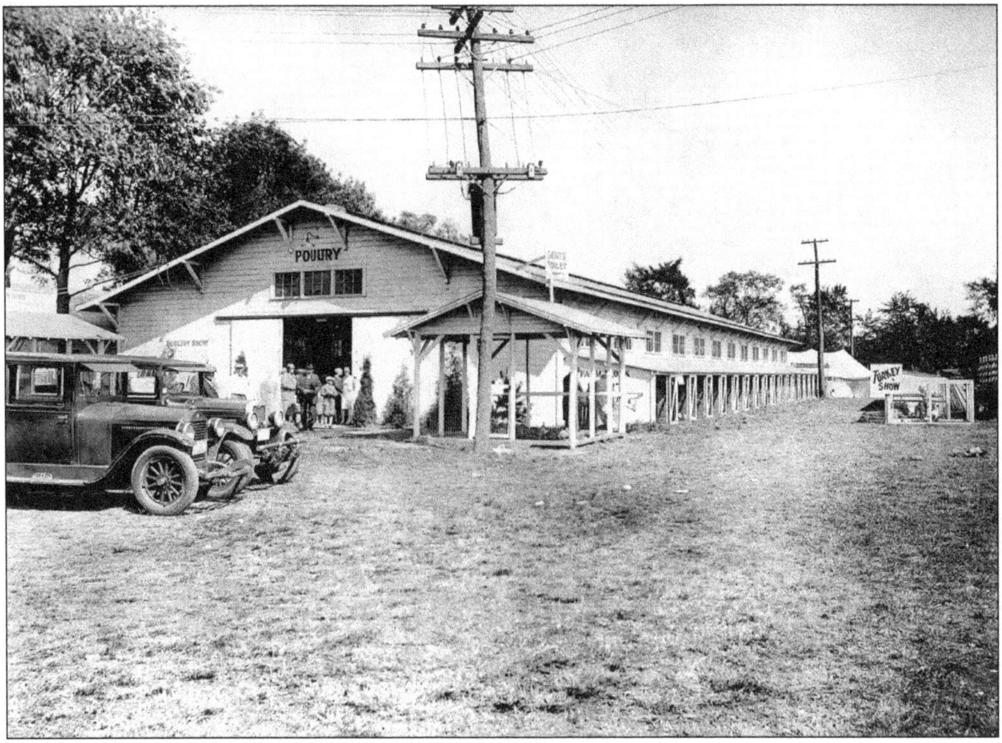

A fabulous photograph from 1929 shows the poultry building with its attached turkey pens. The "Turkey Show" advertised on the large sign to the right must have been a most important event, as can be seen in photographs of several silver loving cup trophies awarded to the various classes for turkeys. Harold T. Fletcher first exhibited his poultry in this building in the 1920s and served as superintendent there for many years. The fine automobile nearest the photographer is a Dodge, and the other is a Chevrolet.

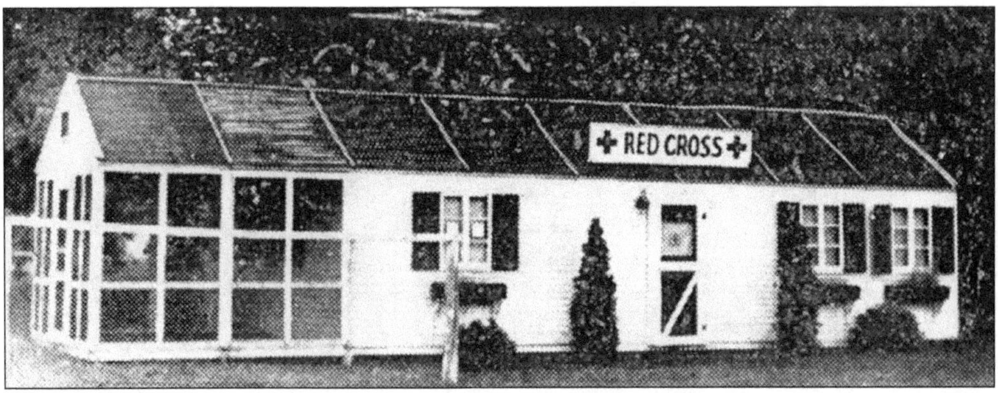

The Women's Welfare and Red Cross Building was erected in 1924 and has served many purposes over the years. The screened porch was removed in recent times to make way for other stands around the center oval. It is partly around the oval on the right-hand side from the main gate where it has stood since 1924. The Women's Building, for home art exhibits, was constructed near the Grange building in 1923 and is now the bee and honey building.

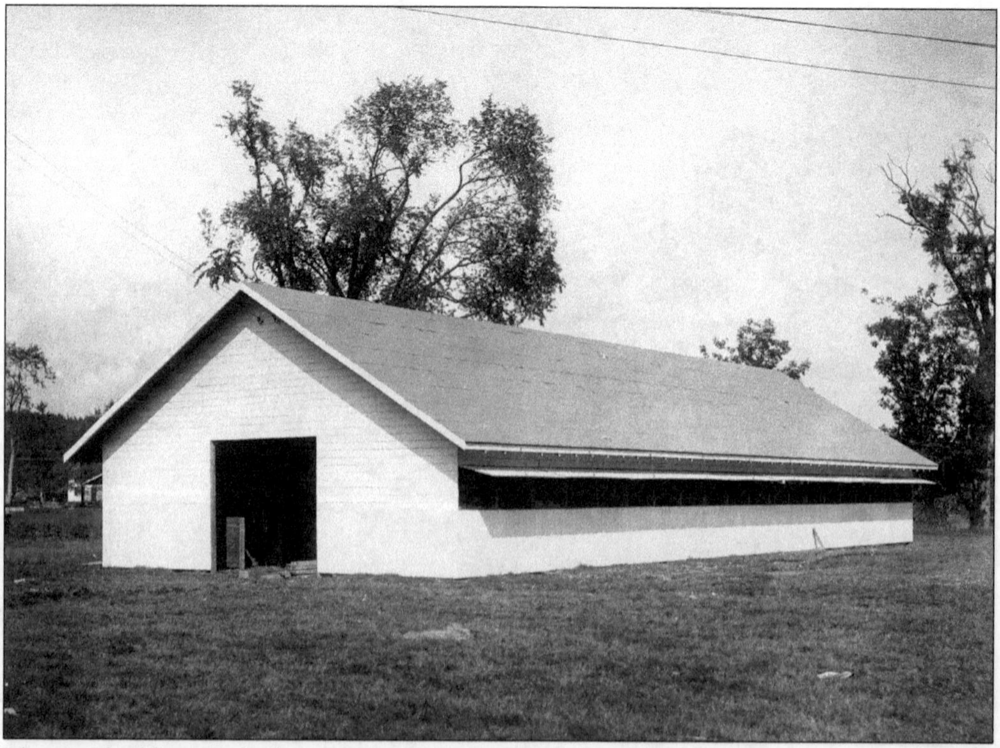

This new building was built in 1921 as a new horse barn and was located approximately where the Coolidge Hall arts building is now located. It was later used for other exhibits when the need for larger buildings were required by the horse exhibitors. Sometime before 1924, a window was installed above the main entrance door.

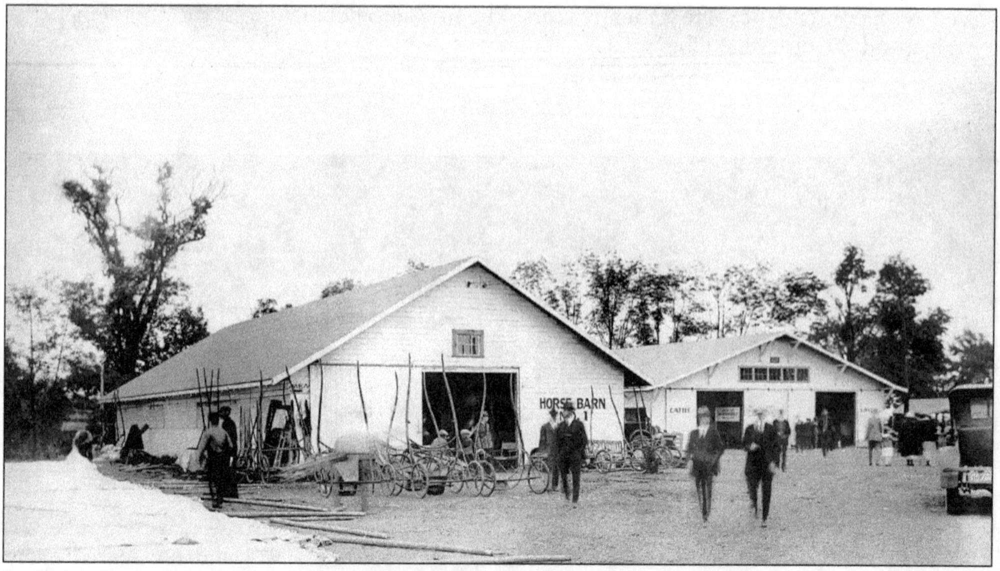

In 1924, horse barn No. 1 was a busy place, as it housed the race horses competing on the racetrack. Note all the sulkeys outside the barn's front entrance. The cattle barn next door housed the show cattle. A sign can be seen advertising the Lawrence Farm Ayrshires from Topsfield inside the building.

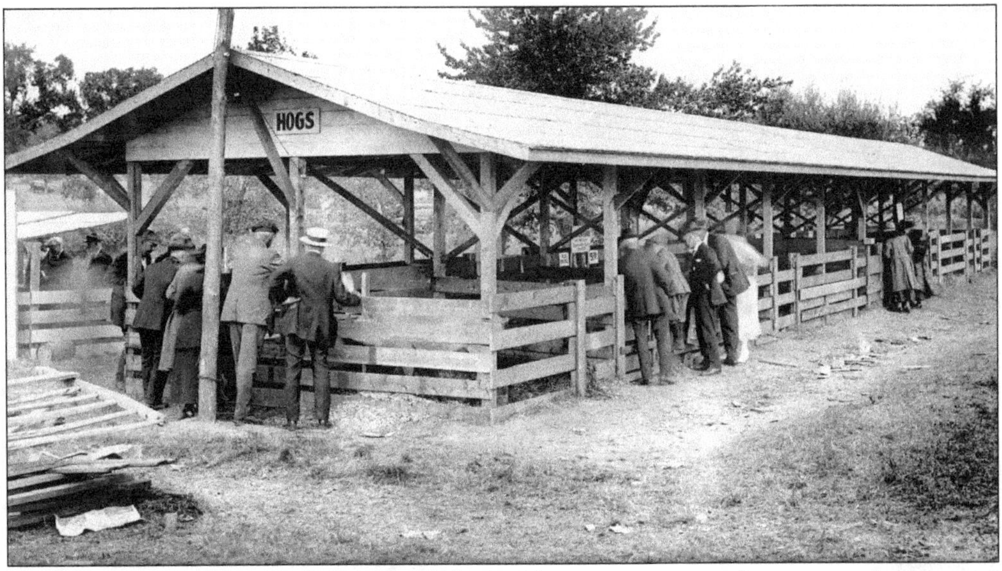

At the back of the fairgrounds near the rabbit building was constructed an open and substantial hog barn. In this 1924 view, men and woman are examining the four-legged inhabitants.

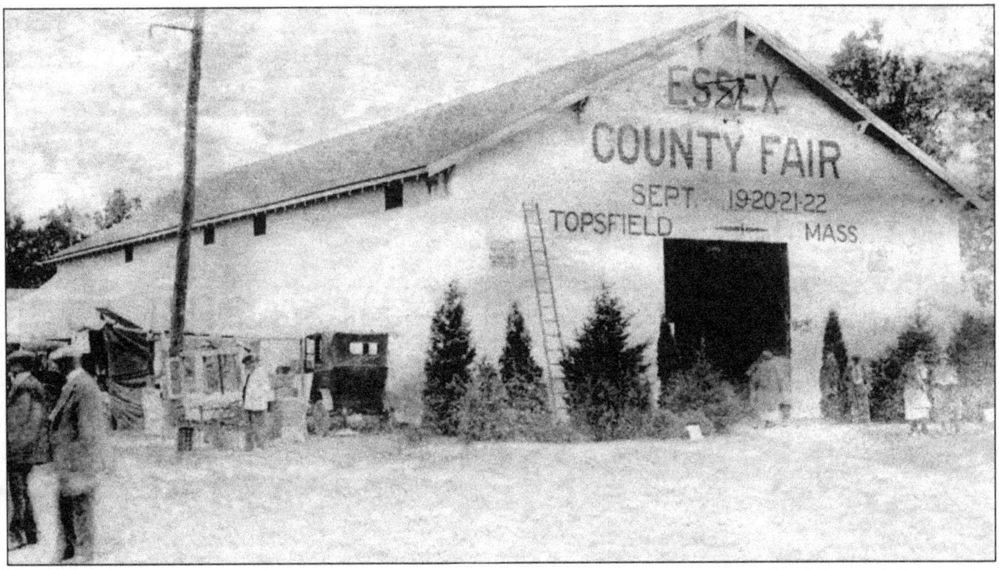

The Grange building was constructed in 1923 and is the only two-story building of this 1920s period ever built on the fairgrounds. The fair dates above the west-end door indicate that this photograph was taken in 1923. The ladder and the lack of casings around the door opening indicate that the building may have just been put up. Note the popcorn vendor and his wagon to the left of the building.

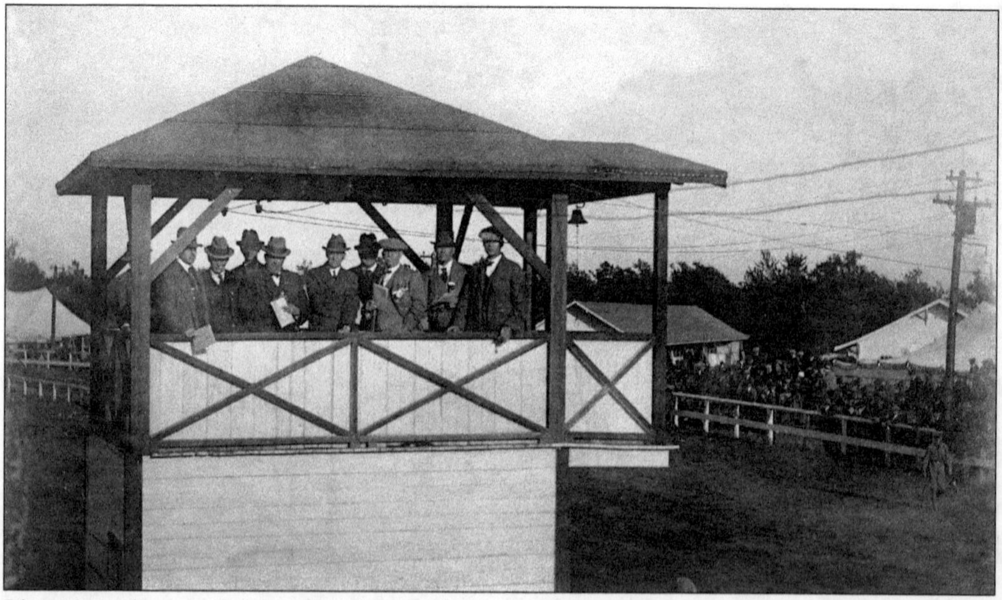

This 1924 closeup of the judges' box at the fair's horse racing track shows a group of judges waiting for the next race to start. The starters bell can be seen hanging on the right end of the box.

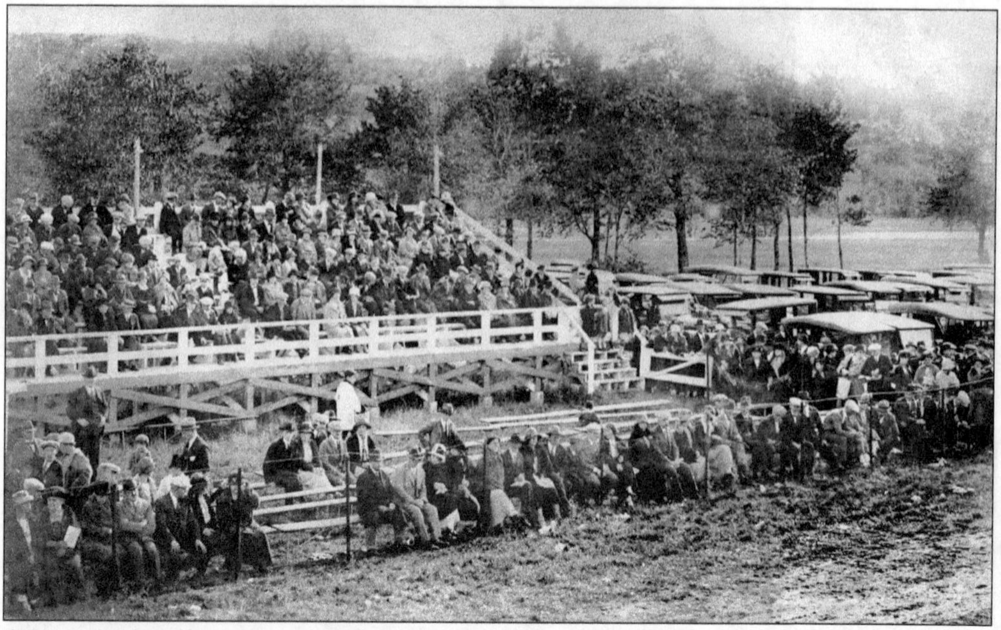

A new grandstand was constructed at the racetrack in 1921 so that race and fair patrons could enjoy the various show performances held on that part of the grounds as well as watch the races. Here, a few years after that construction, we see a large crowd watching the performances, or perhaps they are waiting for a horse race. Note all the soft-top automobiles in the nearby parking lot. There were not many hard-top automobiles manufactured before the late 1920s.

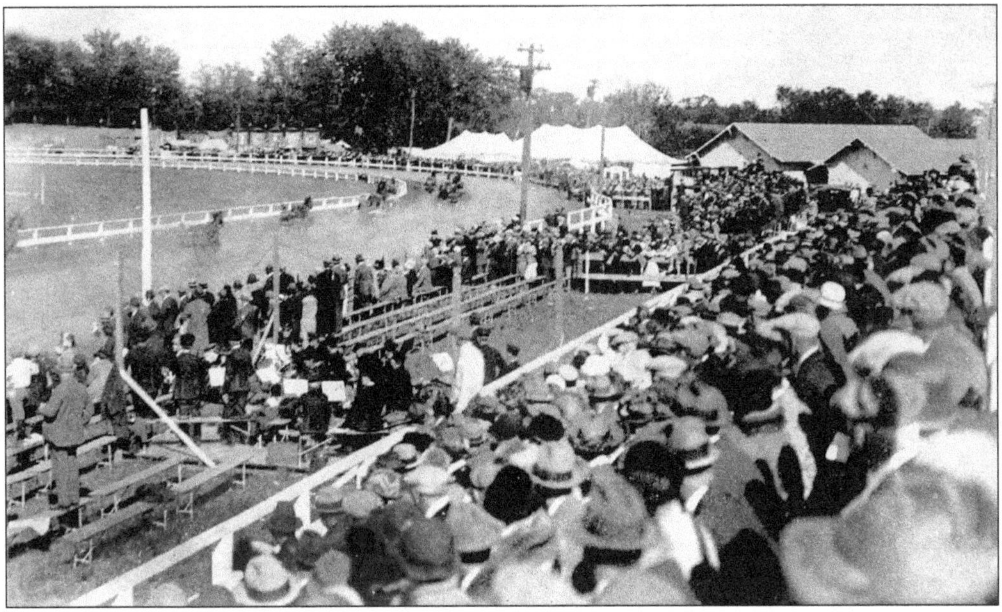

The sulky races have begun at the Topsfield Fair for the 1924 season, and there looks to be a near capacity crowd in the grandstands as well as along the rail. Also flat racing and steeplechases were held at this fair track for many years. Some years, all three forms of horse racing were held during the same fair. For extra thrills, there were even chariot and Roman standing races held at times.

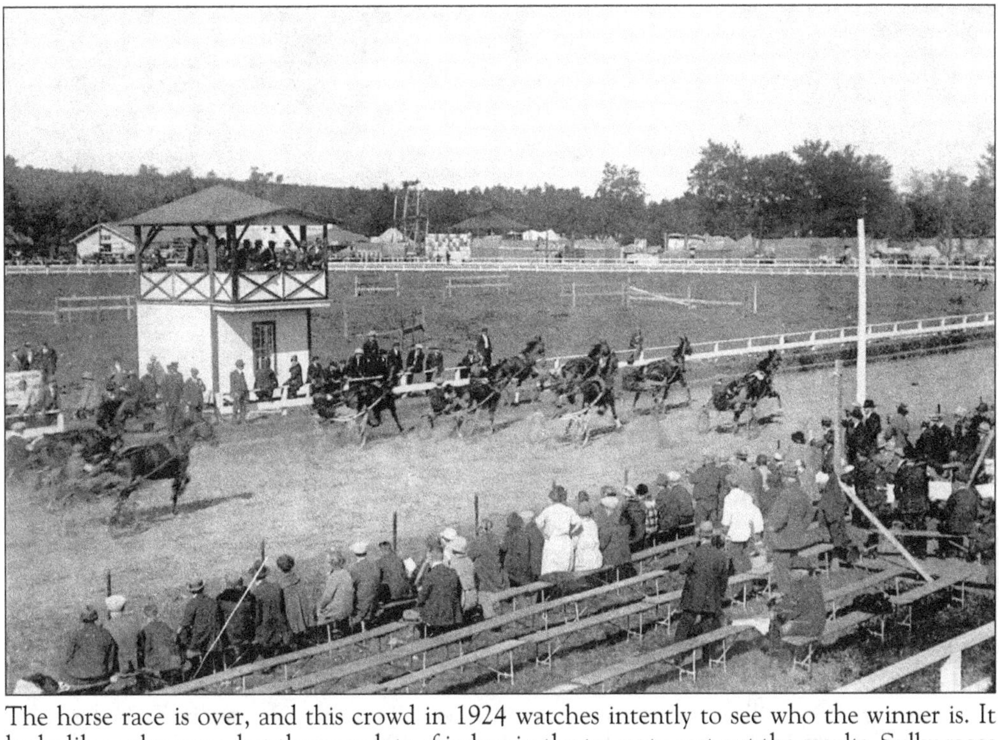

The horse race is over, and this crowd in 1924 watches intently to see who the winner is. It looks like a close one, but there are lots of judges in the tower to sort out the results. Sulky races were very popular for decades at the fairs in New England, and some fairs still have them.

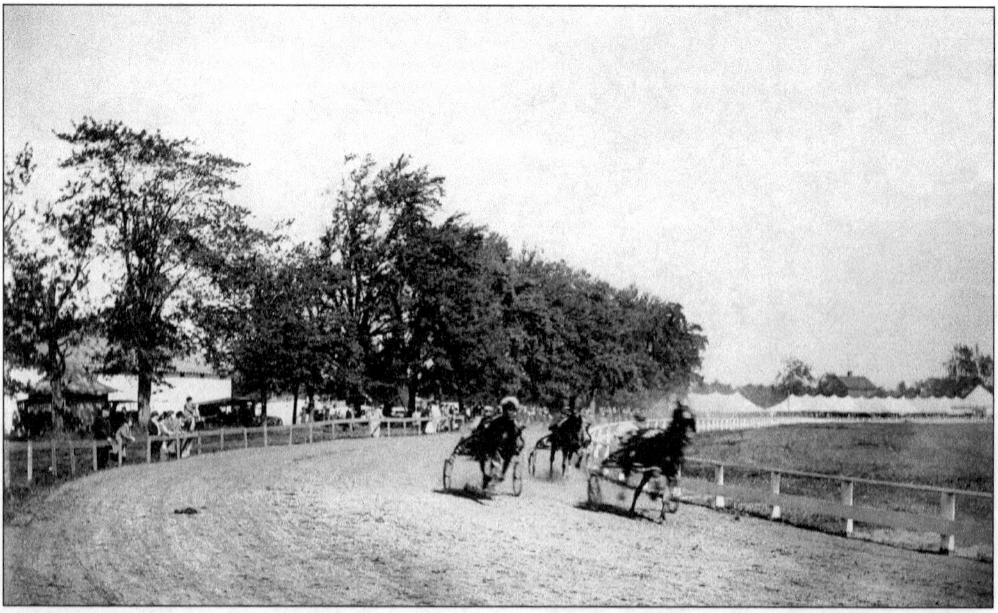

These entrants into the sulky-horse races flash around the track to the delight of the fair's race fans. In the distance are the extensive tents for stabling the race horses and other livestock to be shown at the fair in 1922. Over the years, two other types of horse racing were also concurrently held with the sulky races. The popular flat racing and the steeplechase races were held for many years.

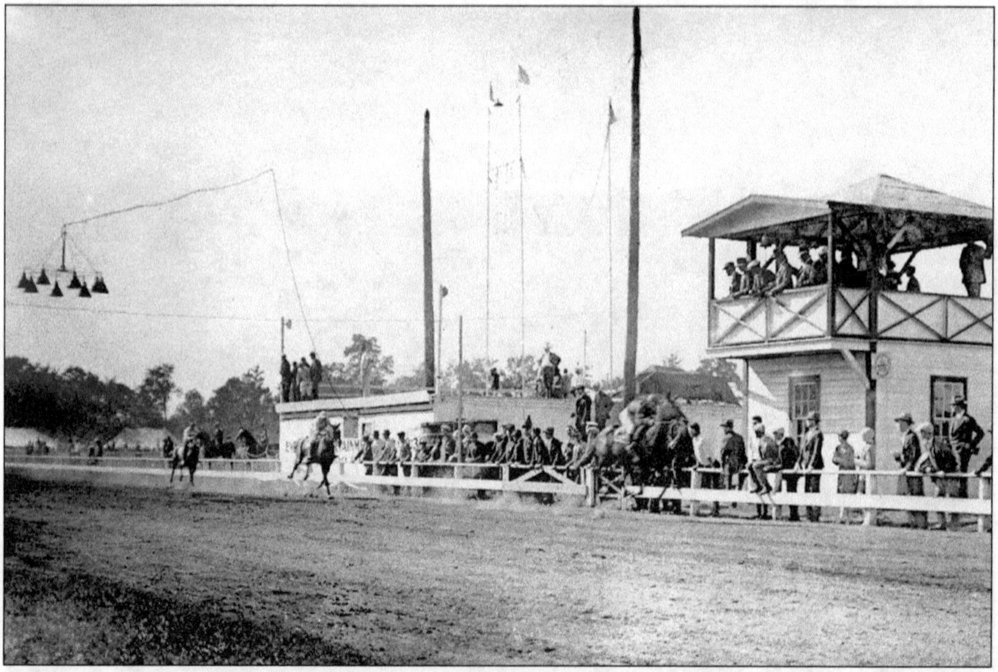

In this 1927 photograph, race horses are just crossing the finish line in front of the judges' booth, and the vaudeville stage is to the left of it, where all the other acts are performed in front of the grandstands. In the early decades of the fair after 1900, many aerialists and other circus-type acts were seen on this stage to amaze the fairgoers daily.

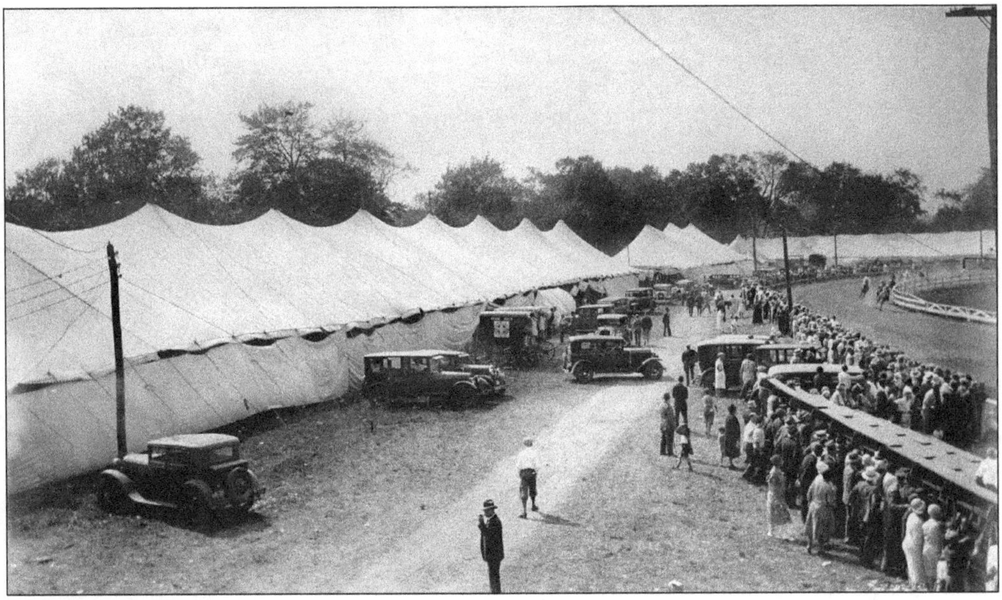

Although the Great Depression was in full swing in 1930, the horse racing events were still extremely popular with the fair's spectators. The horses were stabled in large tents containing box stalls situated along the track. Some fine horseless carriages are parked near ringside as excited patrons hurry to place their bets at the booths (right) before the next race starts. One race can be seen already in progress here.

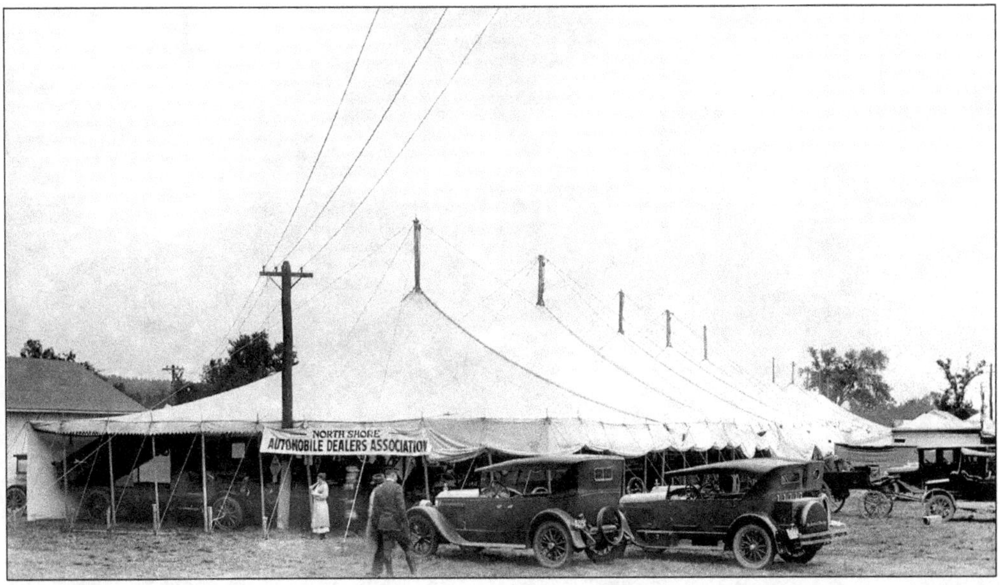

Outside of the North Shore Automobile Dealers Association tent in 1922, we see parked new automobiles, two of which have dealer's number plates on them with the D designation in the number. The closest one is a new Packard. Under the tent to the left is a new Mack dump-bodied truck, with all four tires made of solid rubber. Next to the tent are two milk wagons; one is from the Wason MacDonald Dairy in Haverhill.

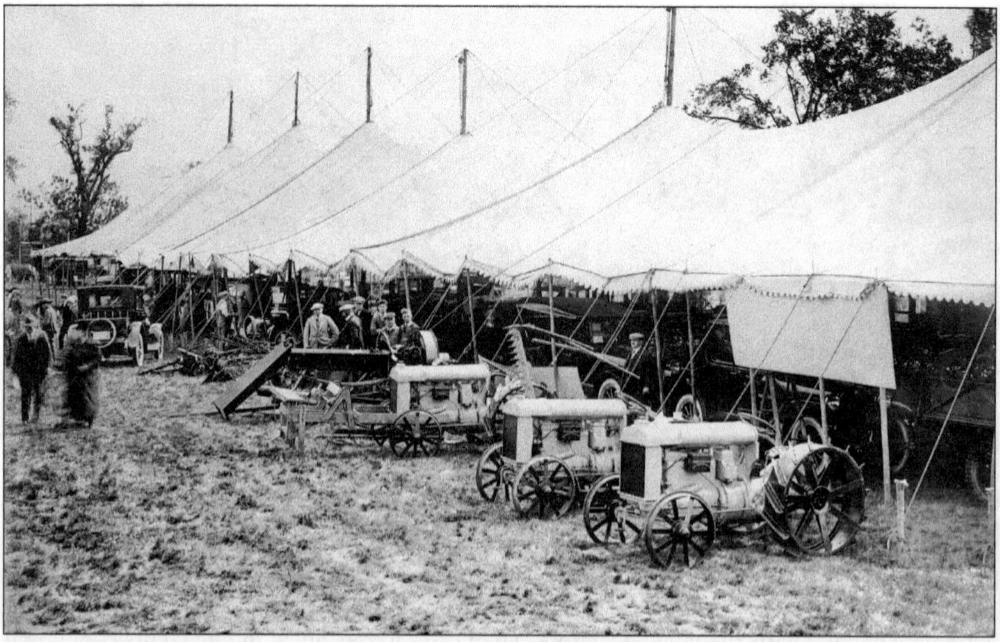

Just outside the automobile dealers tent are several 1922 Fordson tractors along with various implements designed to use power from the tractors. At many earlier agricultural fairs, premiums were awarded for new and improved farm machinery of all kinds, such as mowing machines, rakes, plows, cultivators, and even wagons.

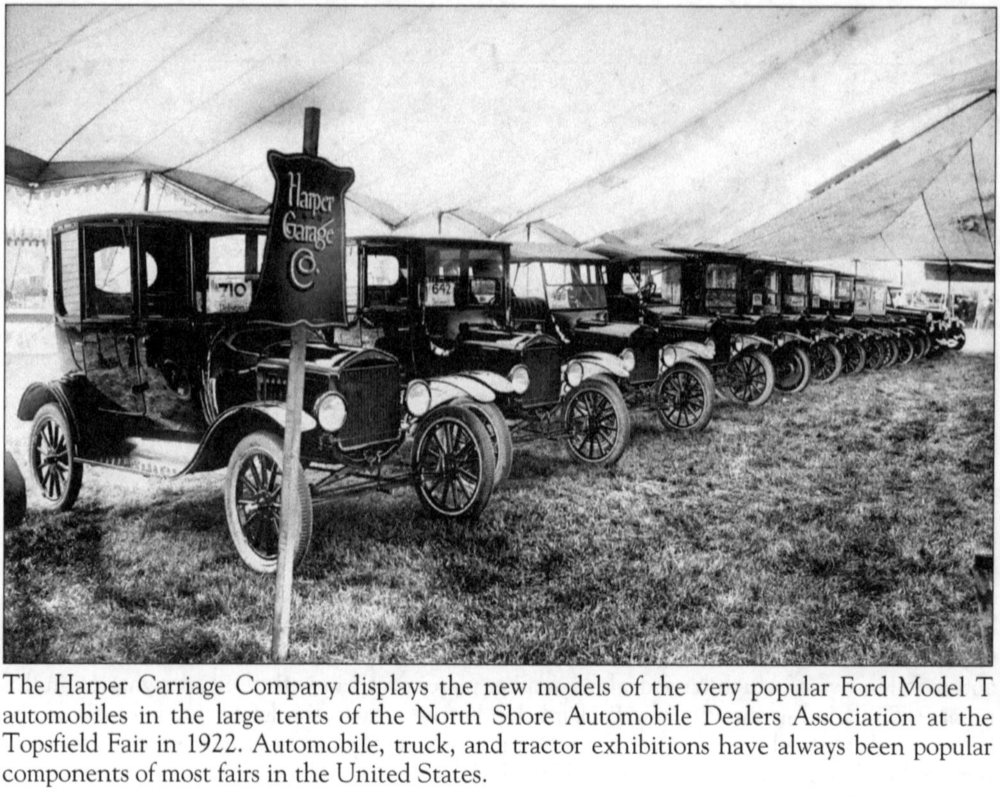

The Harper Carriage Company displays the new models of the very popular Ford Model T automobiles in the large tents of the North Shore Automobile Dealers Association at the Topsfield Fair in 1922. Automobile, truck, and tractor exhibitions have always been popular components of most fairs in the United States.

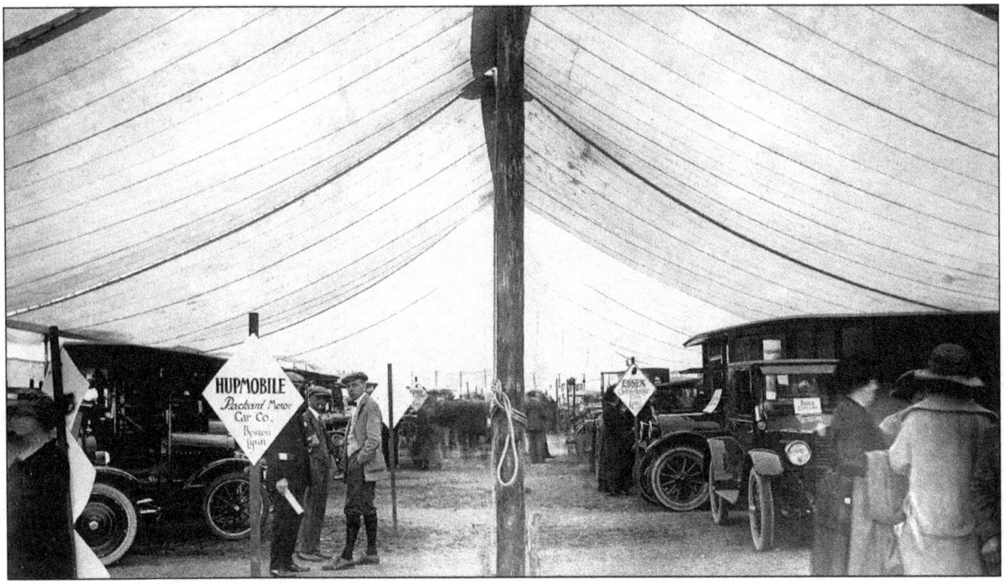

As seen in this 1922 photograph, many new automobiles are crowded into the sales tents of the North Shore Automobile Dealers Association. The Packard Motor Car Company of Lynn and Boston was selling Hupmobiles and Packards. The Oxford Carriage Company of Lynn was selling new Essex models, and the Ford Motor Company had new Fords at the other end of the dealer's tent.

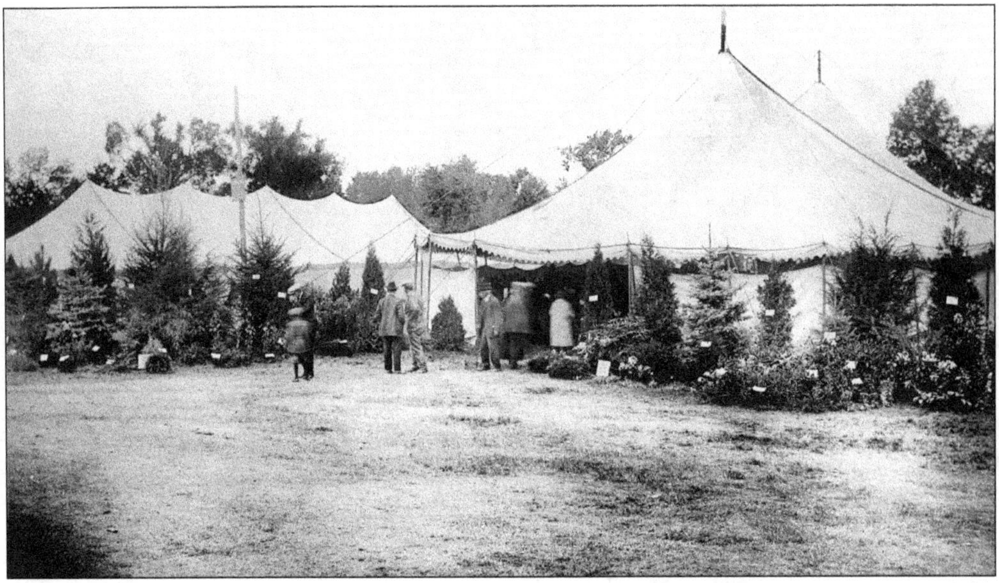

Pictured is another view of the flower show tents at the 1924 Essex County Fair at Topsfield. In front of these tents, many fine examples of trees, shrubs, and flowers have been planted by Cherry Hill Nurseries in West Newbury. The flower show seems to be very well attended by the public, as people are coming and going from these tents.

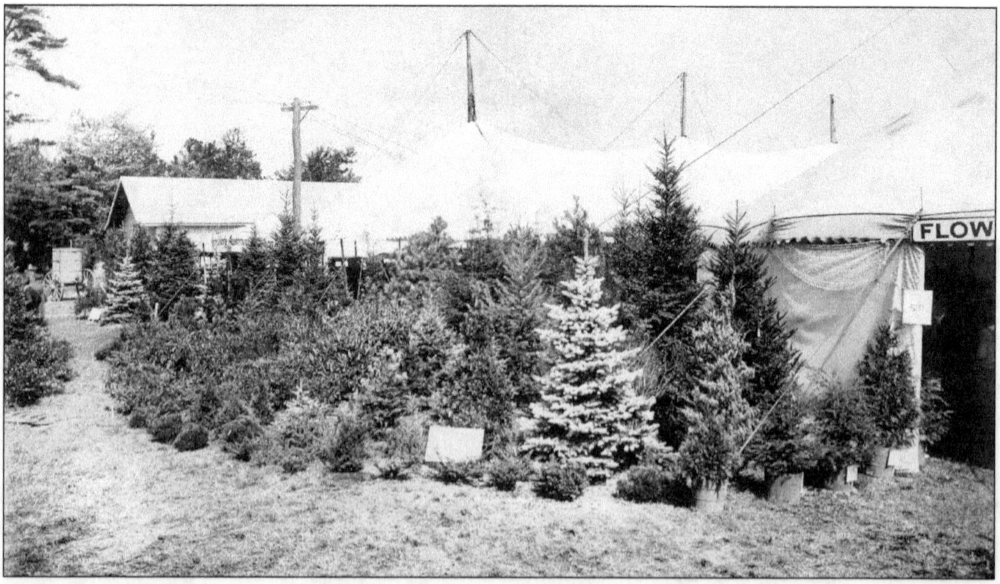

Outside the flower show tent, exhibits of trees and shrubbery have been planted to beautify the fairgrounds and to advertise several of the various nurseries in the county in the early 1920s. The sign in the foreground indicates that this particular planting was put in by the Boxford Highlands Nursery.

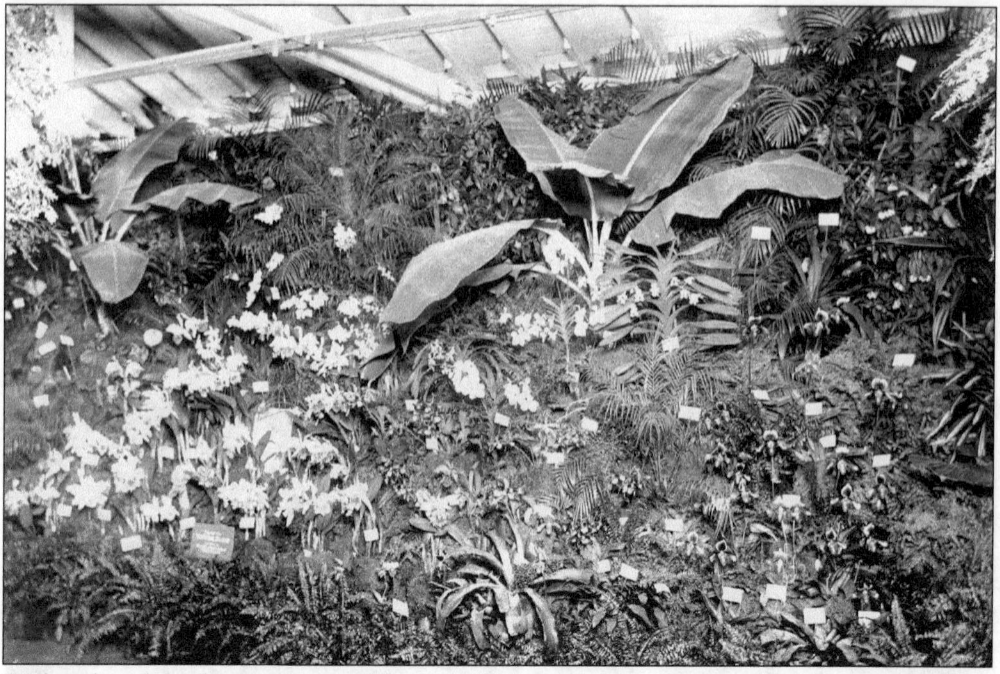

Within the relatively new plant and flower building, we see a very unusual exhibit in September 1929 of tropical plants, predominantly showy flowering orchids flanked by ferns, palms, and two banana trees. Many of the fine estates in the area had large plant conservatories and greenhouses. This superb exhibit was from one of those private estates in Manchester. Note that all of the plants and orchids are carefully named with discreet labels for the public to read. Today's flower shows usually still require exhibitors to label all the plant materials in their displays.

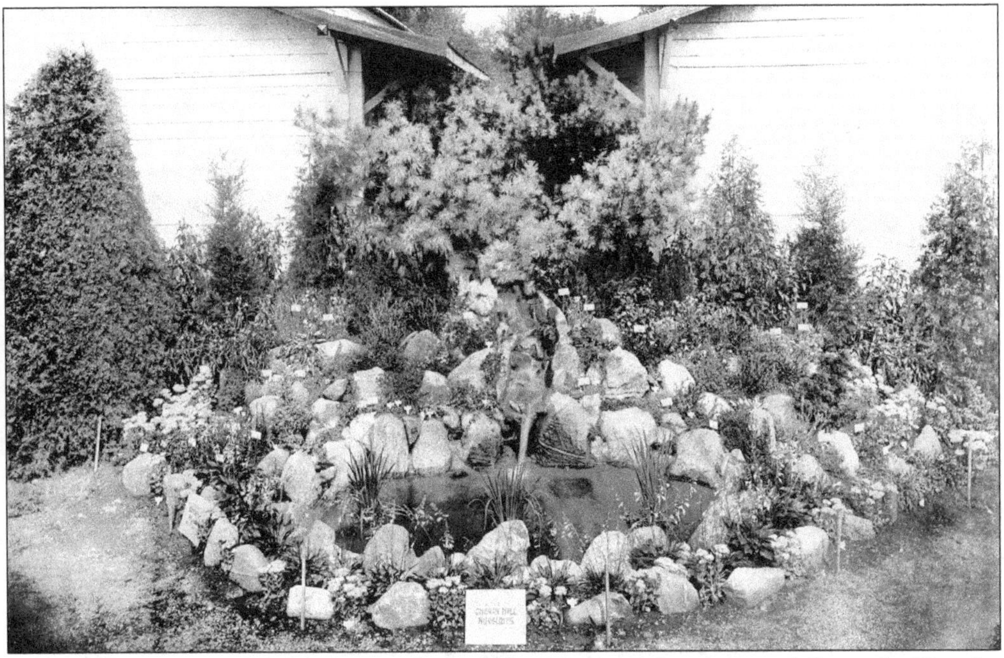

A beautiful water garden has been constructed in front of the plant and flower building by Cherry Hill Nurseries of West Newbury for the 1928 fair. Waterfalls are still very popular in any flower show even today, and there are always several at the Topsfield Fair in the flower building. The sound of water in a garden does something to relax people and make them feel good, so most garden centers and nurseries today sell components for building and operating them, as well as the plant materials for their landscaping.

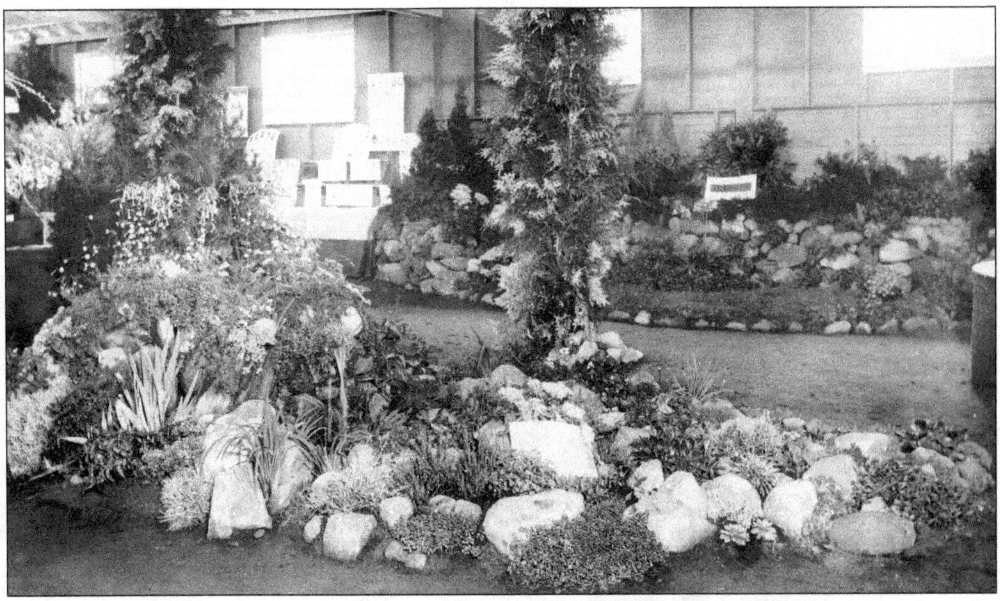

The inside of the flower show building was filled with landscapes and water features such as the waterfall in this exhibit by the White and Johnson Company of Wakefield. The display behind this one was entered by Brecks. A section of the show is always reserved for the exhibition of individual plants and cut flowers by both the amateur and professional gardener as well.

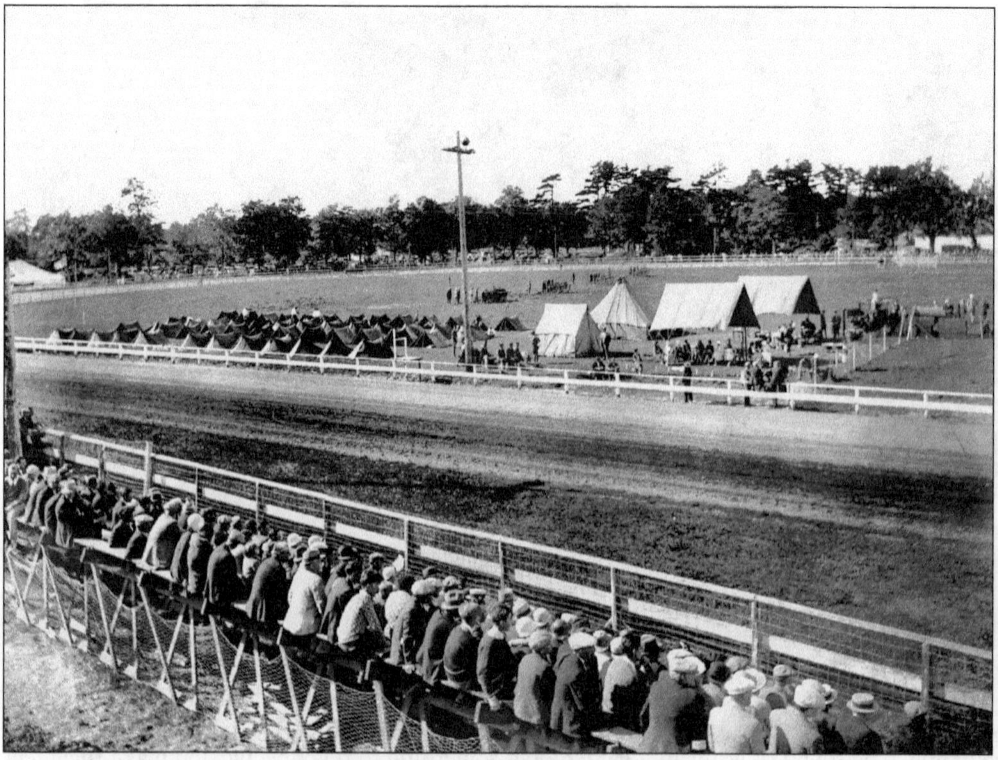

During the 1929 fair, the army had a large encampment in the infield of the racetrack, complete with artillery pieces. There seems to be a keen interest in the encampment as both men and women fair patrons are there in the infield viewing the displays.

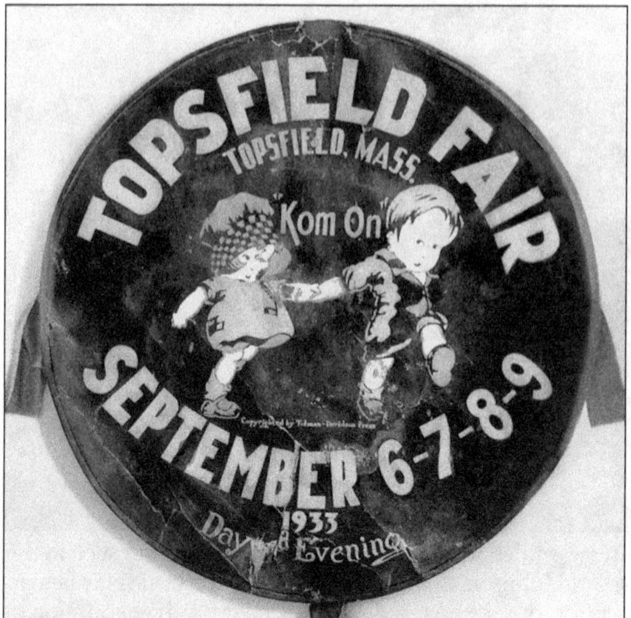

In 1933, an artist or sign painter produced this spare-tire cover for advertising the fair. When the cover was placed over a tire on the rear of a vehicle, it was seen wherever that vehicle traveled—a traveling billboard. A similar advertising banner is shown in another photograph in this volume. This tire cover may be viewed in its frame in the lobby of the Topsfield Fair administration building. (Artifact courtesy of David Walsh of Topsfield.)

The Topsfield Fair archives contain a few very good examples of the colorful exhibitor's gate and parking passes for the fairs held between 1926 and 1930. Exhibitors got specific passes for certain days, as is the practice used even today to admit them onto the fairgrounds.

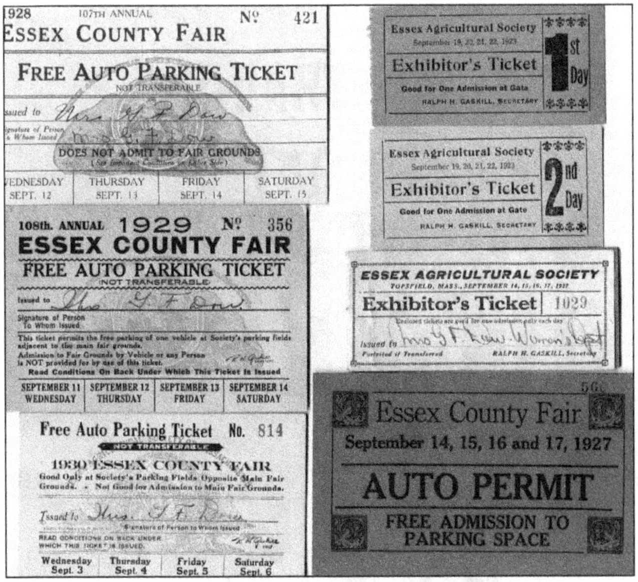

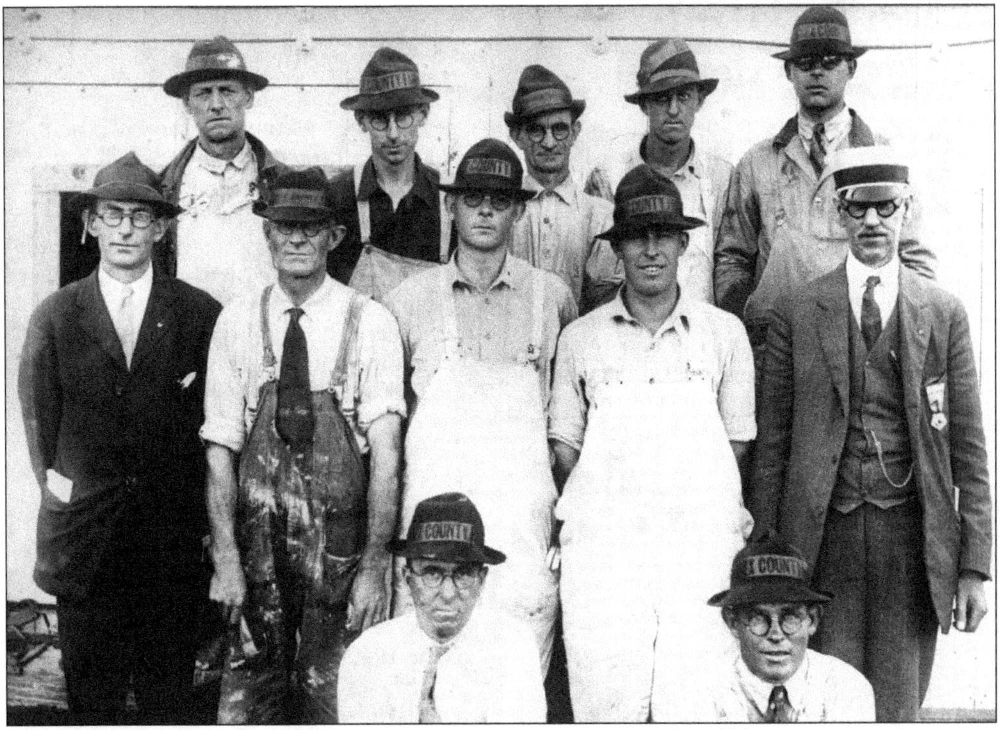

The Essex County Fair work crew is pictured here on September 19, 1927. From left to right are the following: (front row) Walter N. Bemton and John R. Walsh; (middle row) Frank Titus (Harold Fletcher's uncle), an unidentified worker, Stanley Hills, William Wildes, and Eldridge Guilford (a fair official and an employer of many of the men); (back row) John Fuller, an unidentified worker, Mike Coffey, Calvin Watson, and Clarence Castle. This crew had the responsibility of maintaining all of the fair's permanent buildings and the grounds. They would be extremely busy during the fall fair season, when they had to set up many temporary facilities and then clean up the aftermath of the fair and close the buildings for the oncoming winter.

SPECIAL TRAINS
AND
REDUCED RATES
TOPSFIELD FAIR
TOPS THEM ALL
SEPT. 12-13-14-15

Continuous Program = 4 Days and Nights
15 Agricultural Departments
Horse Show and Races
Vaudeville. Fireworks Every Evening

Merchant's and Children's Day, Wed., Sept. 12
ALL CHILDREN ADMITTED FREE

Round Trip Rate 25% Less than Regular

TRAIN SERVICE (Eastern Standard Time)

Round Trip Fares		A.M.	A.M.	P.M.	P.M.	P.M.	P.M.	P.M.	P.M.	P.M.	P.M.
$1.32	Lv. Boston	*8.45	.....	*12.00 / †12.10	¹12.20	*1.30	*3.35	4.25	*5.00	*6.30	*8.30
.86	Lynn	*9.04	.....	*12.19	.....	*1.51	*3.54	.....	*5.21	*6.49	*8.48
.80	Swampscott	*9.08	.....	.....	.....	*1.55	*3.58	.....	.....	*6.53	*8.56
.62	Salem	9.30	11.10	12.40	.....	2.12	4.15	.....	5.51	7.45	9.15
.50	Peabody	9.37	11.17	12.47	.....	2.19	4.22	.....	5.58	7.52	.....
.39	Danversport	9.42	11.22	12.52	.....	2.24	4.27	.....	6.03	7.57	.....
.32	Danvers	9.46	11.26	12.56	.....	2.28	4.31	.....	6.08	8.02	.....
	Ar. Topsfield (Turnpike Crossing)	10.02	11.42	1.12	¹1.23	2.44	4.47	5.32	6.24	8.17	9.45
		A.M.	A.M.	P.M.	P.M.	P.M.	P.M.	P.M.	P.M.	P.M.	P.M.

RETURNING

	A.M.	A.M.	P.M.	P.M.	P.M.	P.M.	P.M.	P.M.
Lv. Topsfield (Turnpike Crossing)	10.10	11.50	1.26	3.25	4.55	7.00	8.25	10.00
Ar. Danvers	.....	.....	.....	3.40	5.13	7.15	8.40	10.15
Danversport	.....	.....	.....	3.42	5.15	7.17	8.42	10.17
Peabody	.....	.....	.....	3.48	5.21	7.23	8.48	10.24
Salem	10.40	12.29	1.56	3.55	5.28	7.30	8.55	10.32
Swampscott	.....	*12.42	.....	*4.27	*6.13	*7.57	*9.11	10.39
Lynn	*11.10	*12.47	*2.27	*4.33	*6.18	*8.02	*9.15	10.43
Ar. Boston	*11.28	*1.05	*2.45	*5.01	*6.36	*8.20	*9.36	11.06
	A.M.	P.M.	P.M.	P.M.	P.M.	P.M.	P.M.	P.M.

Train leaving Topsfield at 10.00 p. m. will stop to leave at Chelsea and Everett.

*Change at Salem. ¹Sat. Only.

Children Between 5 and 12 Years of Age, Half Fare

ASK YOUR AGENT

BOSTON AND MAINE RAILROAD

W. O. WRIGHT
General Passenger Agent

Allen Shaw Print, Beverly, Mass.

The Boston and Maine Railroad operated special passenger trains in the 1920s and 1930s from Boston to the Topsfield Fair. This handbill, one of three in the society's archives, is from 1929 and lists the last stop as the Turnpike Crossing. It was apparently only a wooden open platform on the south side of the tracks near the present bottled gas dealer. The train tracks were put in behind the fairgrounds in 1854. By 1934, the Boston and Maine was sending out buses from Boston to carry passengers to the fair according to the fair's programs. There must have been a reduction in passenger service to Topsfield as some passenger trains continued coming to the town until 1950, when the passenger service was finally terminated. A fairly complete set of annual reports from 1820 to the 1940s can be found in the archives of the Essex Agricultural Society. The reports are a great source of the detailed history of the fair for each year, including the exhibit lists and the premiums won by named exhibitors, with hometowns, in each department and class, as well as the other activities of the society, including membership and officer lists. The Essex Agricultural Society is looking to find *Transactions* for the 1820s and 1830s.

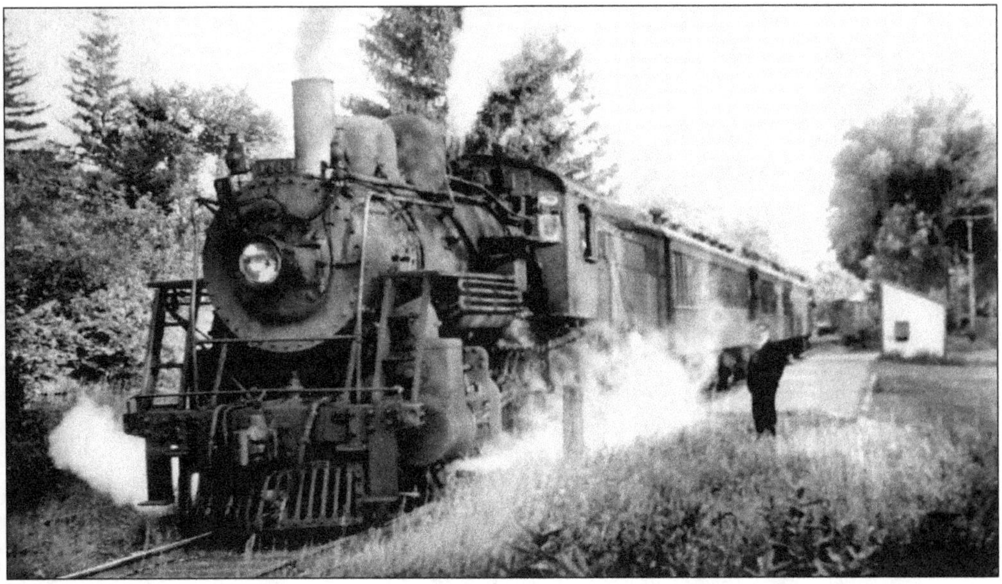

Passenger trains such as this one, shown at a small passenger shelter in Topsfield in the late 1930s, brought people to the fair during the 1920s and early 1930s. The Boston and Maine Railroad provided special and regular train service for fairgoers, stopping at the Turnpike Crossing platform to let them detrain close to the fair's main gate. After the train discharged its fair-going passengers, it continued to Georgetown to turn on the wye there and return to Boston again. (Photograph courtesy of the Daybreak Cafe, Topsfield.)

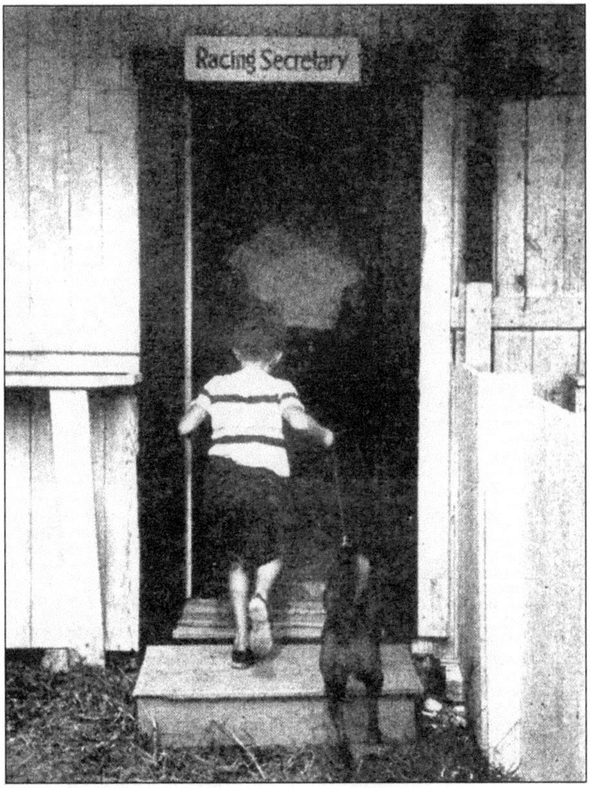

Johnathan Bond of Topsfield was eight years old in 1935, when this picture was taken. He is seen entering the race secretary's office at the first Topsfield Fair "Mutt Derby." Johnathan did not know exactly what breed his dog Penny was, but he finally listed her as a "four-way" dog. The race secretary's building is still on the fairgrounds today, but the dog racing track is long gone. (Photograph courtesy of Johnathan Bond.)

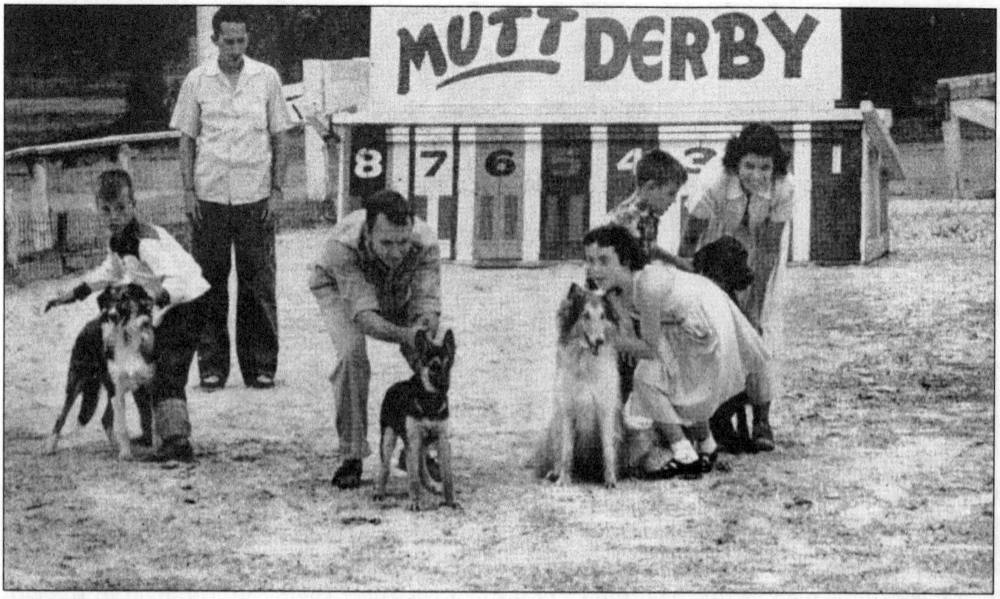

When this mutt race starts, the collie promenades, a puppy gets pushed, the Great Dane lies down and grins at his owner, and the beagle charges down the track after the rabbit. (Photograph courtesy of Johnathan Bond.)

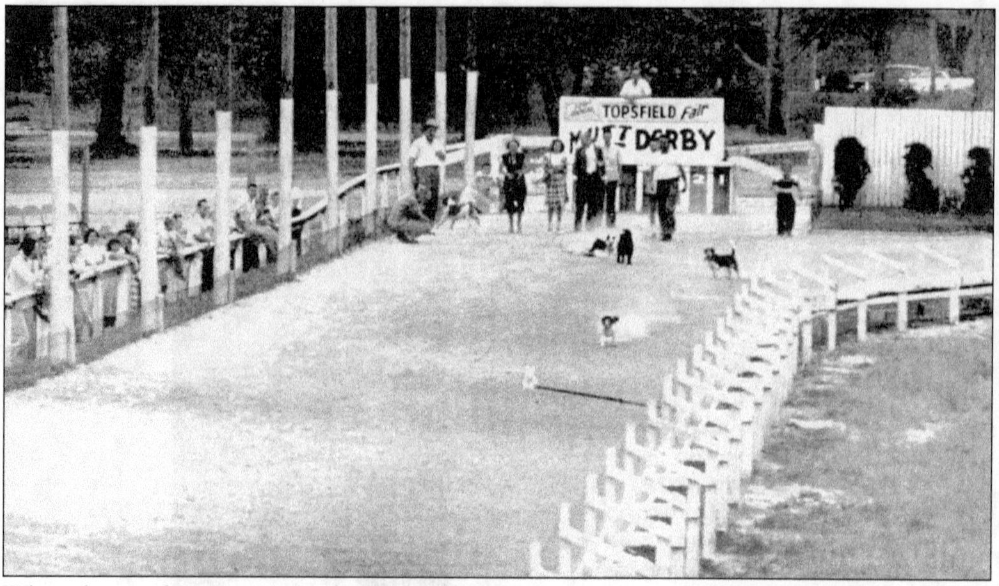

When the two-horsepower mechanical rabbit Topsy slid seductively past the dogs in the start line and the starter shouted "Go," most of the dogs did not go, as they just sat there wagging their tails and licking their embarrassed owners' faces. The little beagle was the only dog to run after the rabbit in the group. After all, beagles were bred to chase rabbits, so his owner had some chance of success in the derby. The races mostly turned into dogfights as excited canines ran into other unfamiliar participants. Of the 12 races held that day, with 34 dogs entered, only 6 dogs crossed the finish line. (Photograph courtesy of Johnathan Bond.)

Shown is a recent interior photograph of the flower show building with a display of chrysanthemums being used to decorate the gardens in the process of being built before the fair officially opens. The U.S. flag remained the centerpiece of the show for that fair in 2001. (Photograph courtesy of John Deanna Cole.)

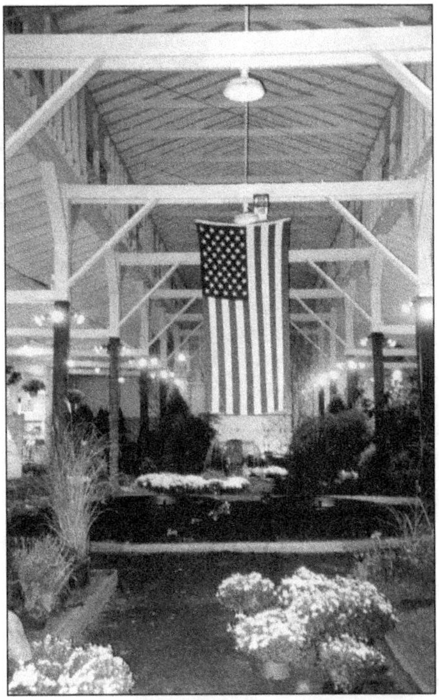

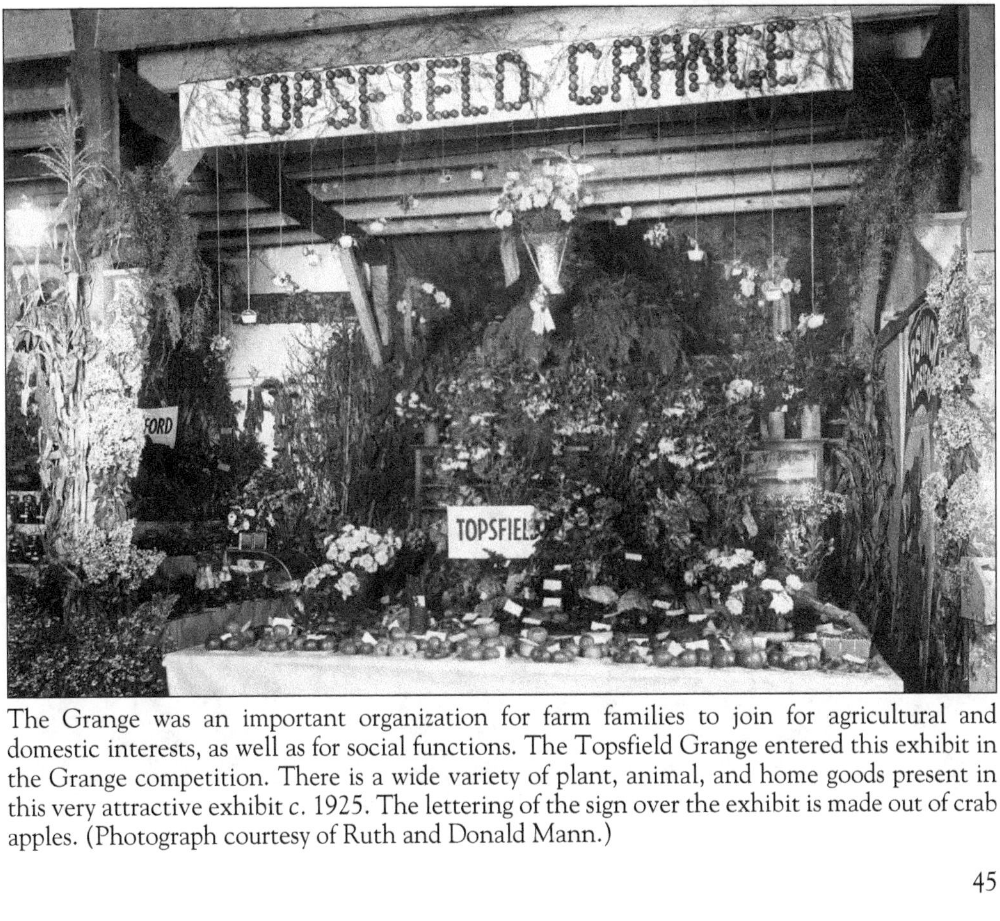

The Grange was an important organization for farm families to join for agricultural and domestic interests, as well as for social functions. The Topsfield Grange entered this exhibit in the Grange competition. There is a wide variety of plant, animal, and home goods present in this very attractive exhibit c. 1925. The lettering of the sign over the exhibit is made out of crab apples. (Photograph courtesy of Ruth and Donald Mann.)

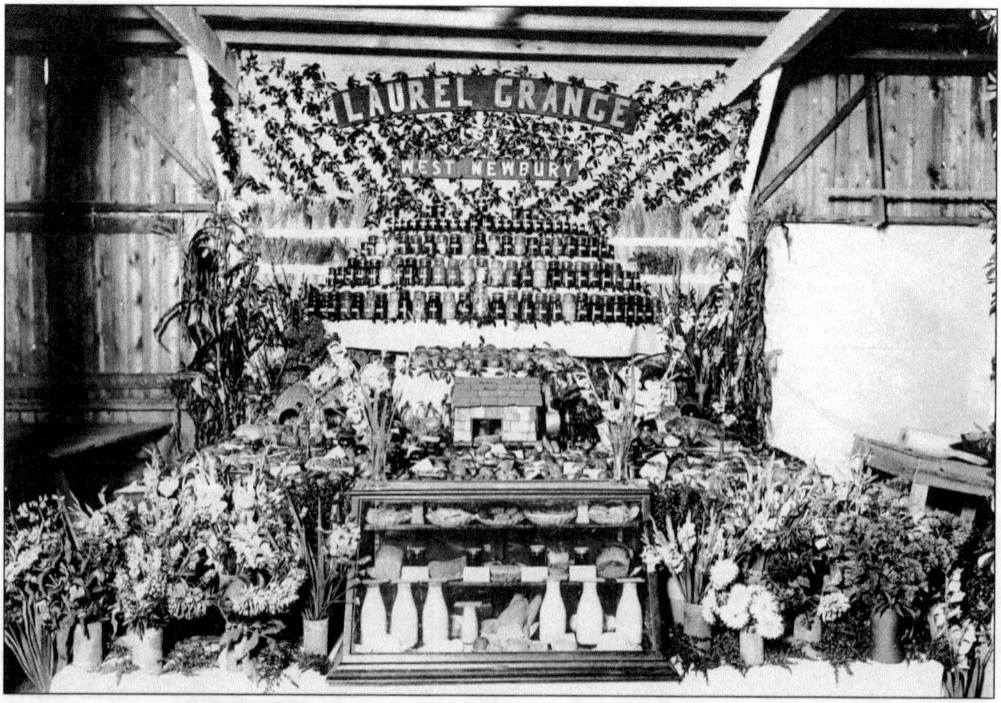

The West Newbury Laurel Grange entered the Grange exhibit contest with a most impressive display of farm produce and canning preserves. The competition in those days must have been very keen, as many Granges participated in the fair, thus making it very difficult for the judges to decide the winners.

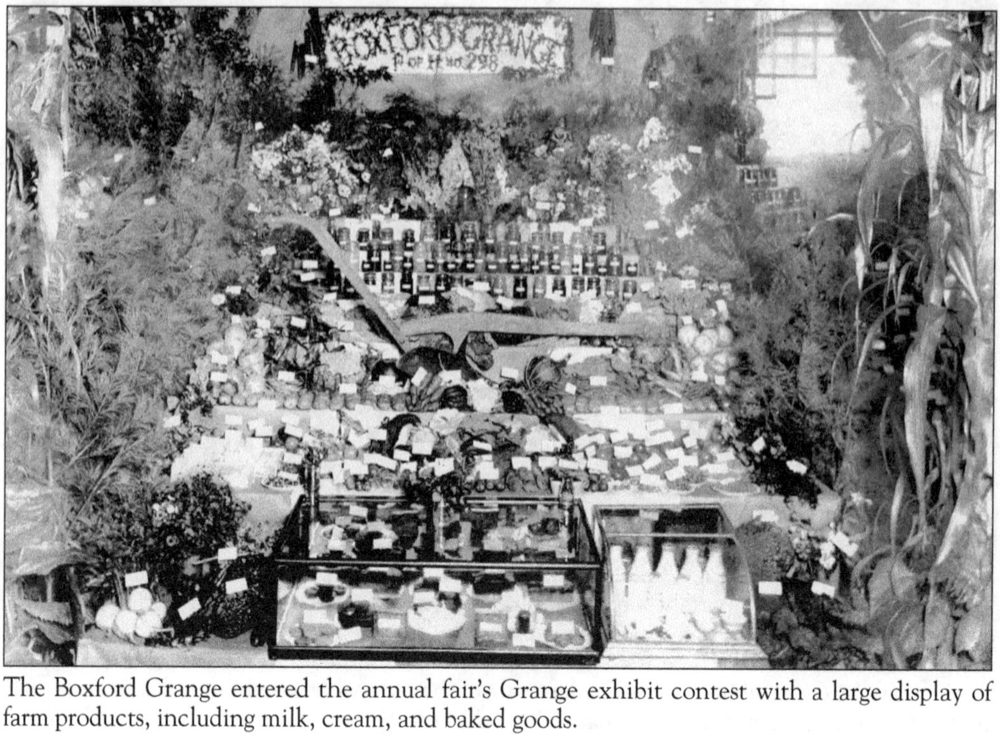

The Boxford Grange entered the annual fair's Grange exhibit contest with a large display of farm products, including milk, cream, and baked goods.

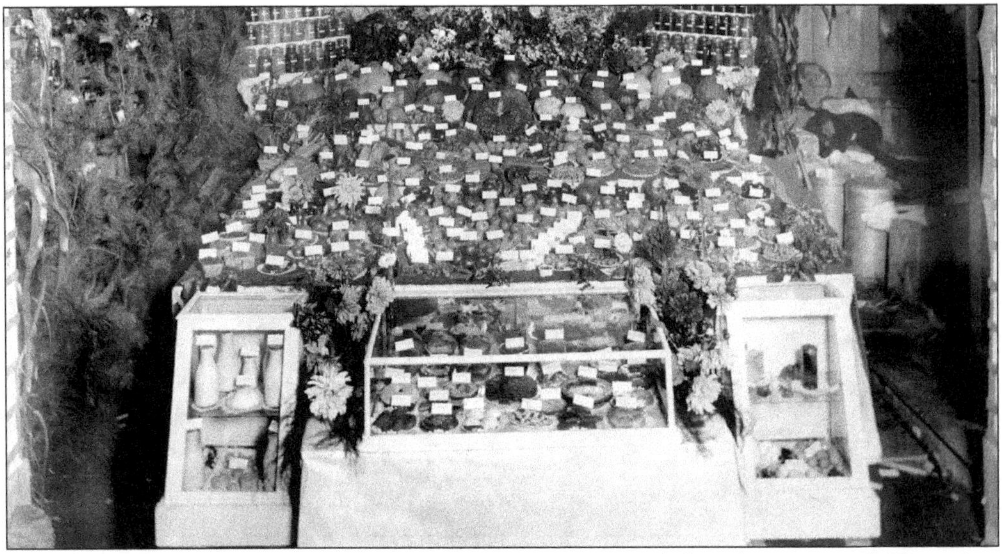

The Grange building was built in 1923 to house the Grange exhibits then, as it still does today. There were many more Granges in earlier years than today as people have moved away from the family farms. The 1928 Beverly Grange, Patrons of Husbandry, No. 306 exhibit (shown here) was well organized with all manner of flowers and agricultural produce, even cream and milk in bottles and homemade foods in glass jars. Some exhibits even displayed sewing handiwork and craft items made by Grange members.

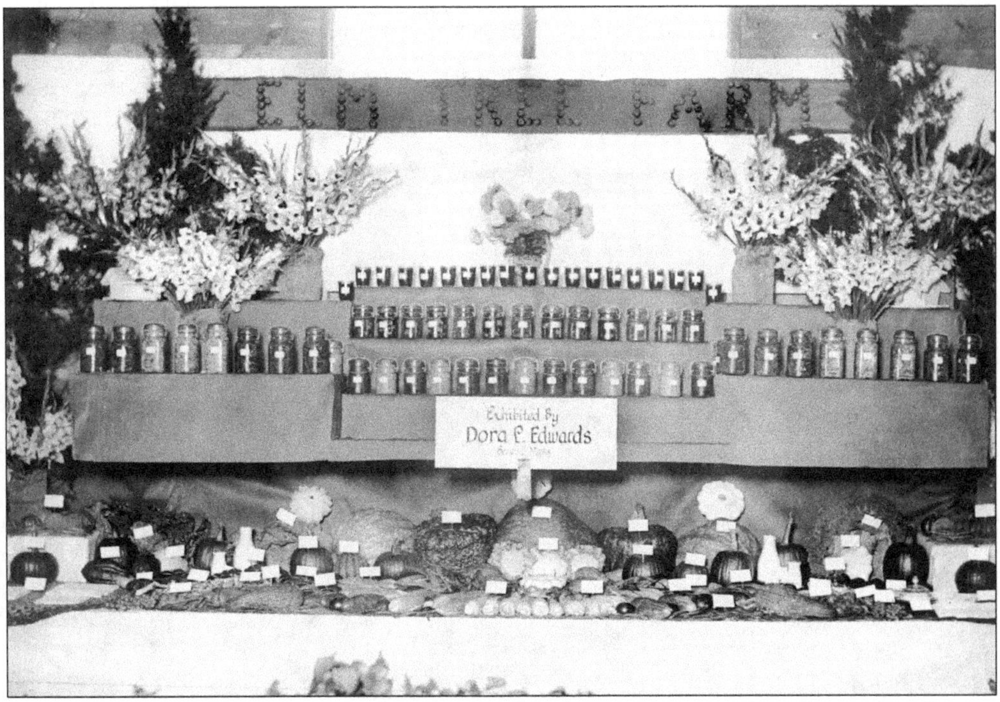

We are not sure just who Dora L. Edwards from the Elm Tree Farm in Beverly was, but she was certainly hardworking and ambitious if her 1927 Topsfield Fair exhibit of produce and canning is any indication of her diligence.

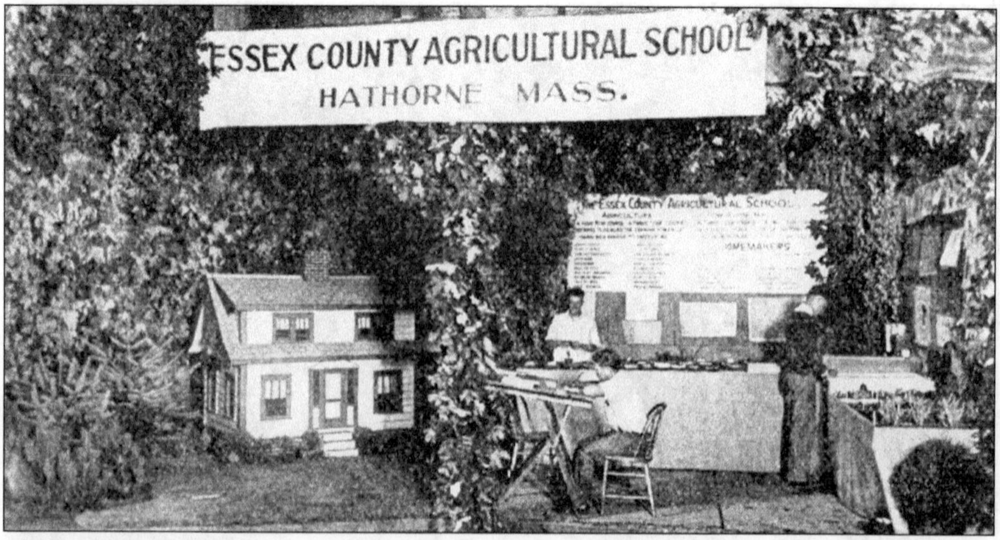

The Essex Agricultural School, or "Essex Aggie," in nearby Hathorne, was founded in 1913 as a result of the Smith-Hughs Act, which provided federal funds for setting up agricultural education and homemaking programs for students at the high-school level. Massachusetts set up three of the schools at the county level in Essex, Bristol, and Norfolk Counties, and one was set up in North Hampton as the Smith Agricultural School. All the other secondary agricultural programs were only departments in regular high schools across the nation. Beginning in 1914, the students regularly exhibited at the Topsfield Fair. Here, in 1934, we see one of those exhibits in which agriculture and homemaking were emphasized. The Essex Agricultural and Technical Institute (as it now is called) still exhibits. Students and FFA members show their animals and plants at each annual fair in Topsfield.

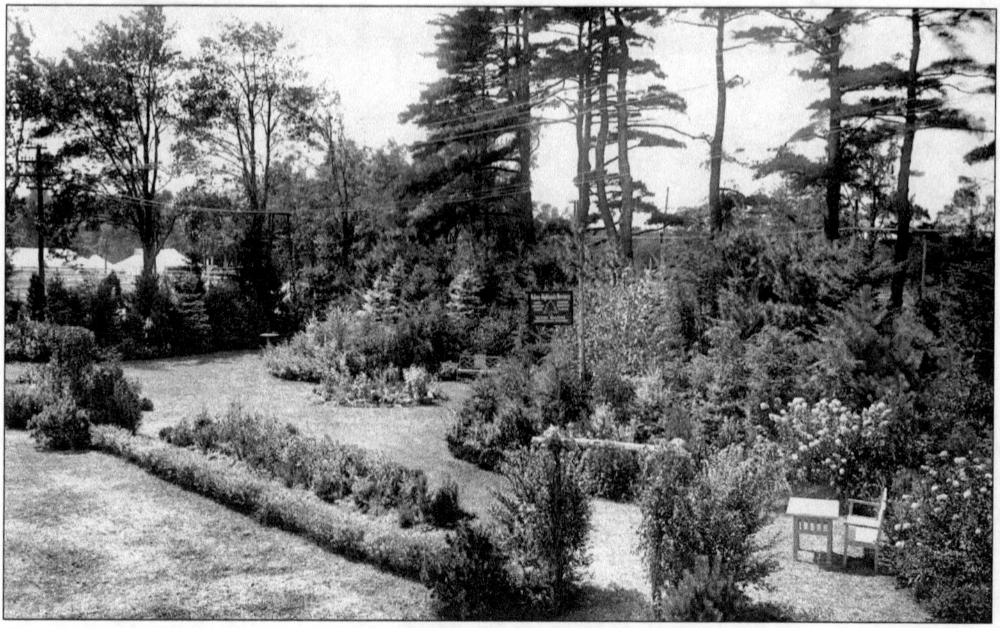

Kelsey-Highland Nursery, in nearby East Boxford, planned and executed this restful area near the turnpike (Route 1), shown here in 1929. In later years, the Joy Town Kiddie Park was put here among the trees, shrubs, and flowers.

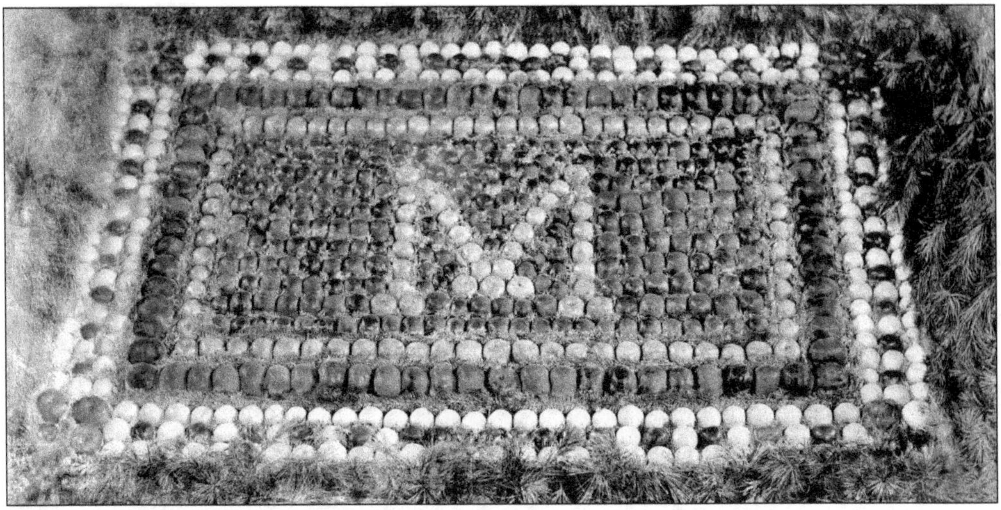

The fruit building in 1924 at the Topsfield Fair was graced by a large pictorial display made from a variety of different colored apples. The display had a large letter M in the center and was exhibited by Charles Warren Mann. The Charles W. Mann Farm is now known as Mann Orchards. The orchards are still in the Mann family and are operated by Charles's granddaughter Ruth Mann Fitzgerald with her husband and son. Charles Mann joined the Essex Agricultural Society in the mid-1880s and later became a trustee of the society. (Photograph courtesy of Ruth and Donald Mann.)

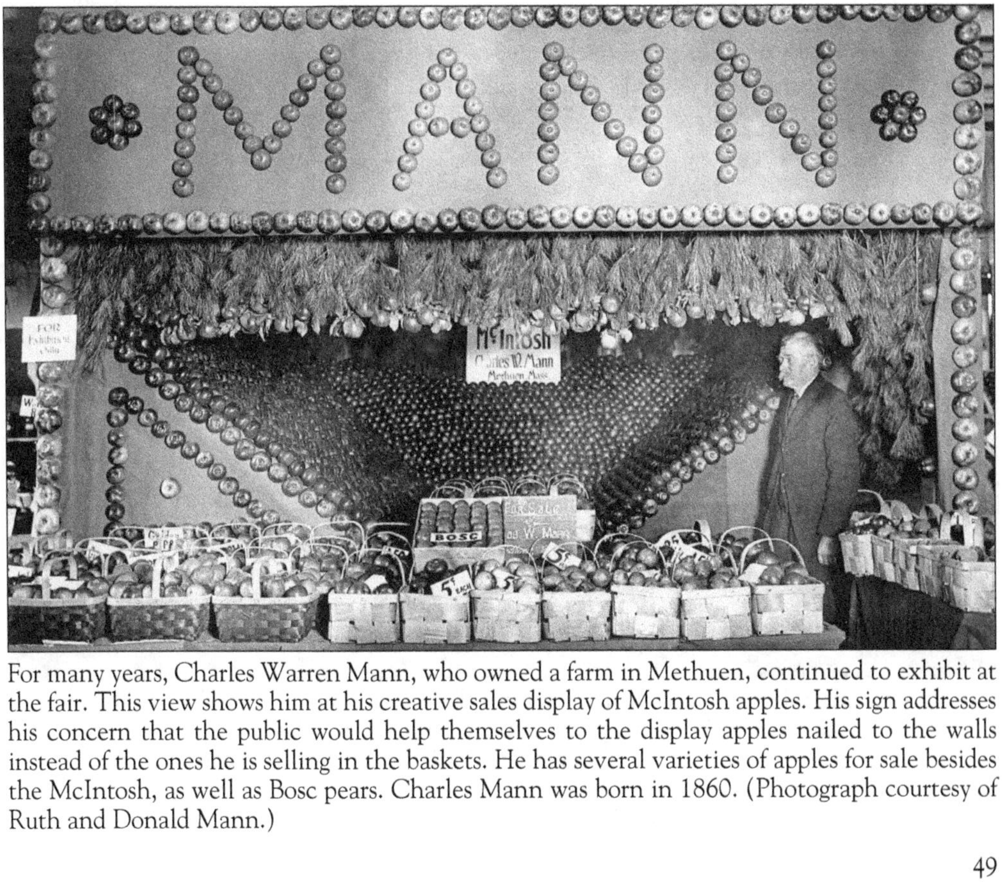

For many years, Charles Warren Mann, who owned a farm in Methuen, continued to exhibit at the fair. This view shows him at his creative sales display of McIntosh apples. His sign addresses his concern that the public would help themselves to the display apples nailed to the walls instead of the ones he is selling in the baskets. He has several varieties of apples for sale besides the McIntosh, as well as Bosc pears. Charles Mann was born in 1860. (Photograph courtesy of Ruth and Donald Mann.)

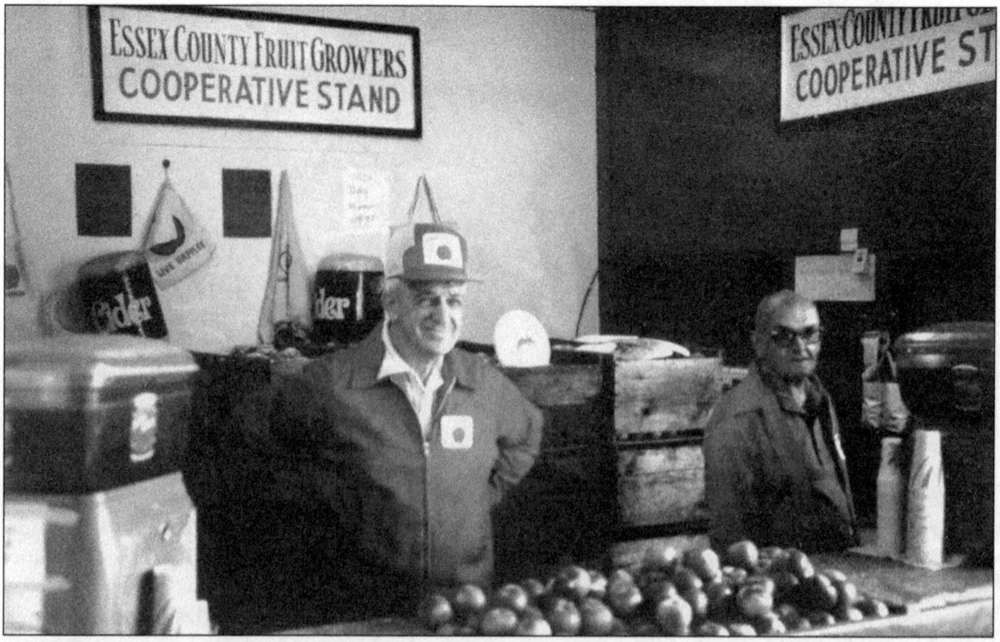

The Essex County Fruit Growers Association has been selling apples and cider in the fruit and vegetable building for countless fairs. Harold "Bud" Kelley and Edward Dwyer, both from Danvers, manned the association's booth with other members from 1975 to 1982. Kelley first worked with Brooksby Farm in Peabody when he returned from World War II in 1946 and later became an instructor and fruit department head at the Essex Agricultural School in Hathorne. (Photograph courtesy of Harold Kelley.)

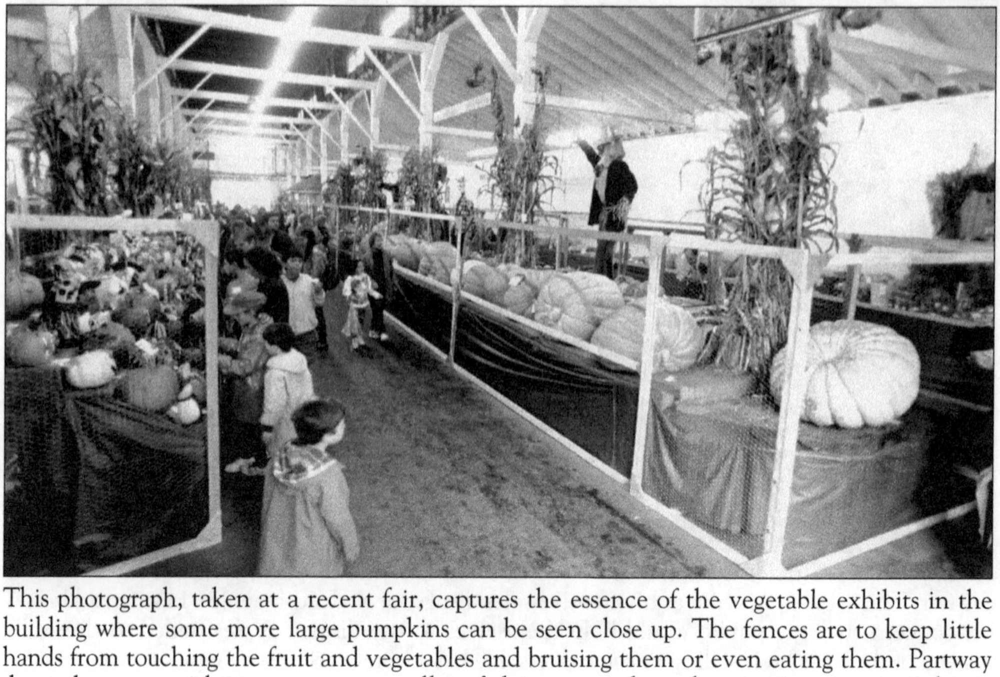

This photograph, taken at a recent fair, captures the essence of the vegetable exhibits in the building where some more large pumpkins can be seen close up. The fences are to keep little hands from touching the fruit and vegetables and bruising them or even eating them. Partway down the center aisle, one can see a small roofed structure where the winning giant pumpkin is displayed in an enclosed case all week for fair patrons to marvel at.

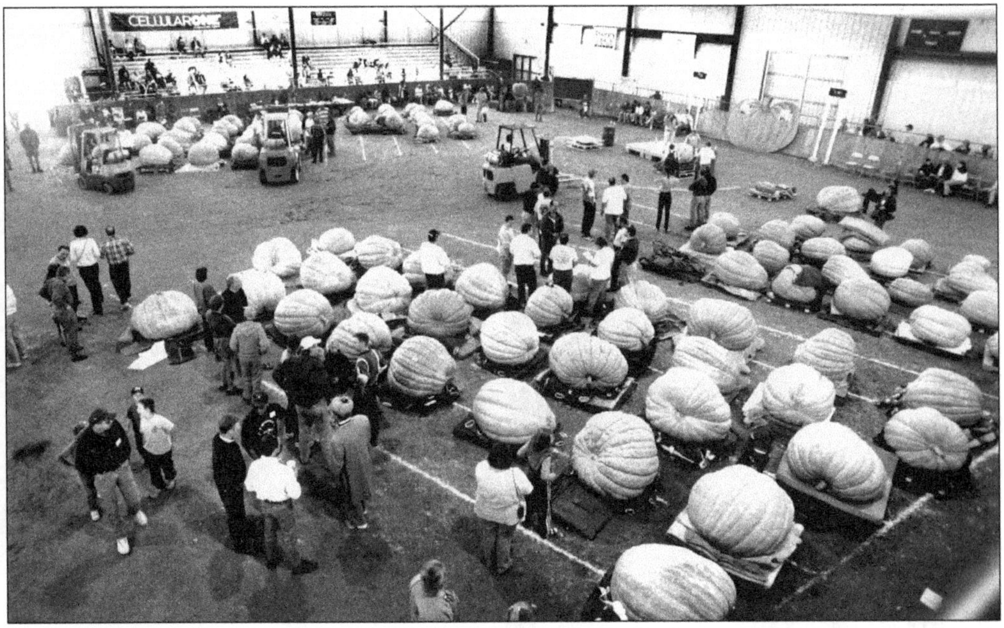

At the pumpkin contest (held on the opening day of the fair), there is great excitement as each pumpkin is gently lifted onto the official scales for weighing. The 2002 winner was Charlie Houghton of new Boston, New Hampshire. His giant pumpkin weighed in at more than 1,337 pounds. Less than 10 years ago, few would believe a pumpkin could ever grow to even 500 pounds. If you look closely at this photograph, perhaps you can spot the winner somewhere in the arena.

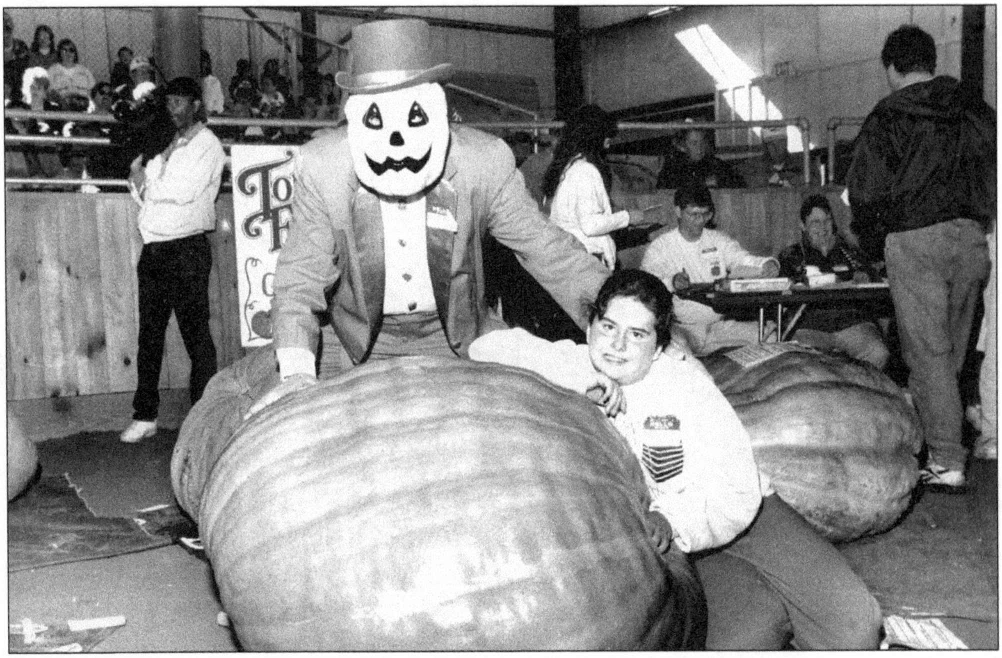

During the pumpkin weigh-off (held in the livestock arena), pumpkin committee member and former contest winner Wayne Hackney of New Milford, Connecticut, usually wears some sort of a pumpkin costume. In this 1995 view, he is wearing a giant pumpkin head. In other years, he may wear his bright-orange suit with a green bow tie.

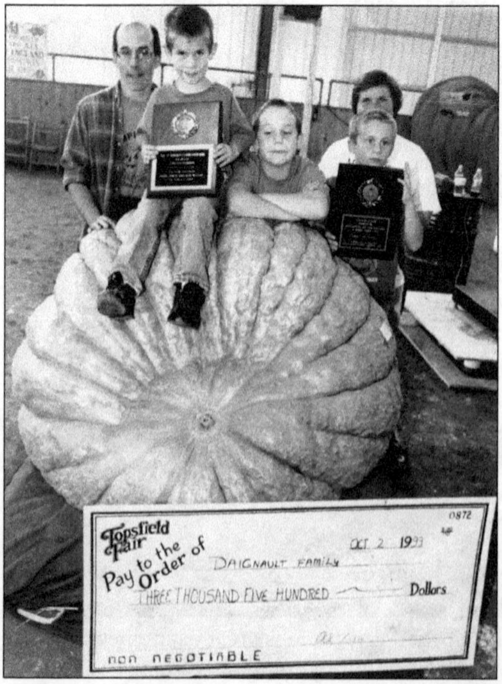

The Daignault family poses here with their $3,000 prize check and trophies for their winning giant pumpkin at the weigh-in on October 2, 1999. One of the trophy plaques is for the first pumpkin ever to exceed 1,000 pounds. It would certainly make a whopper of a jack-o'-lantern at Halloween.

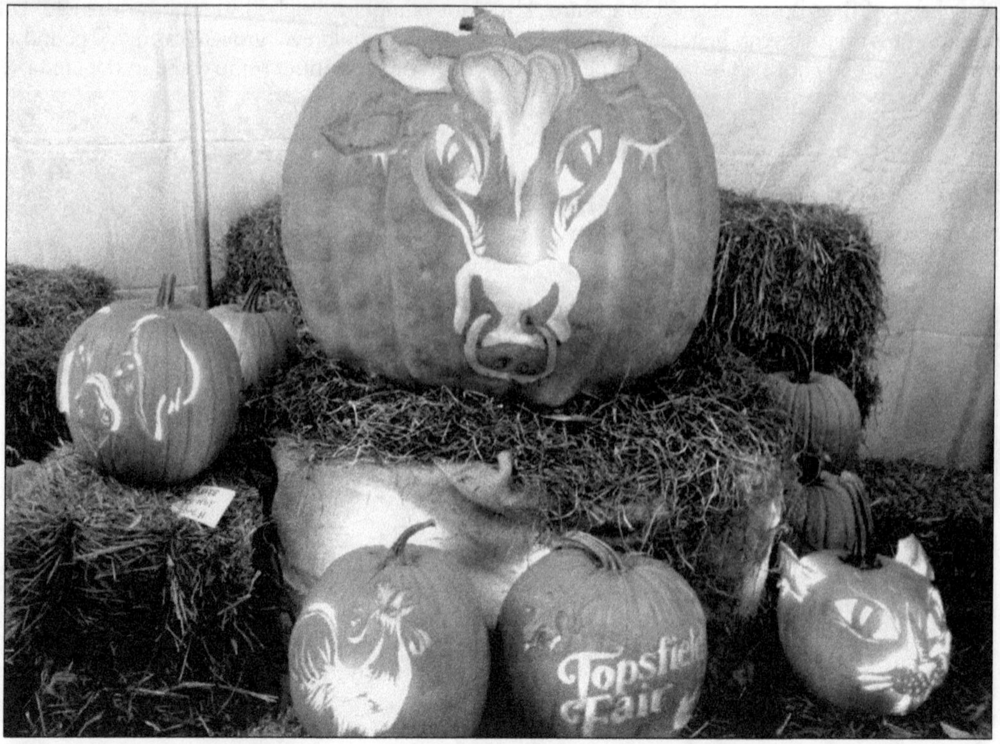

Hugh McMahon, a master carver of jack-o'-lanterns, turns pumpkins into examples of fine art as he thins the walls to varying thicknesses to allow partial light to show through, thus giving the image a shaded effect. His works of art are displayed in the show tent at the back of the main fairgrounds.

In the industrial building, there were many commercial exhibits. This photograph shows the attractively decorated booth that several electric-appliance dealers pooled their resources for to showcase the new aluminum Maytag electric washing machines.

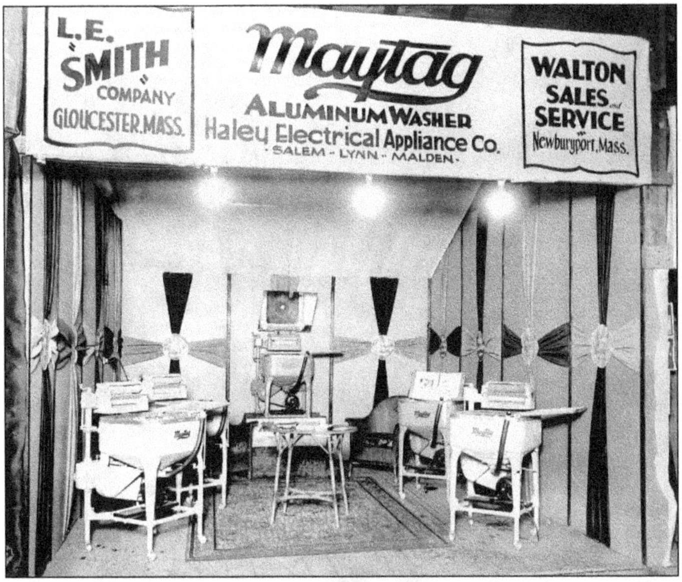

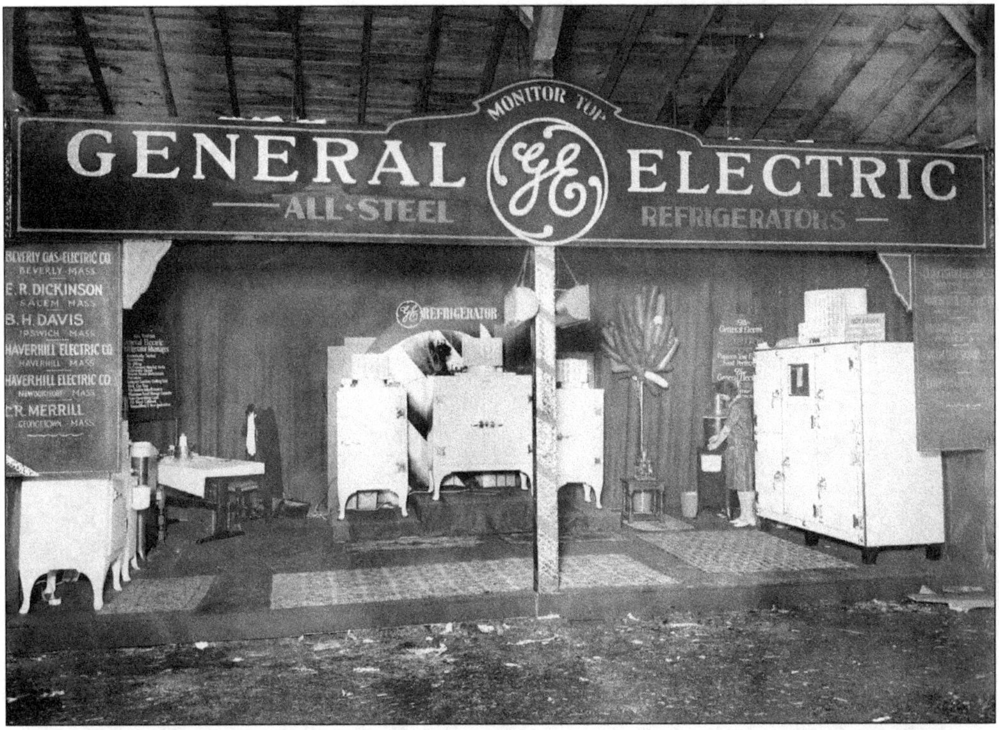

Many local businesses, such as the collective of electric and appliance companies listed on the sign on the left of the booth, joined together to exhibit the latest in monitor top electric refrigerators in the industrial building in 1930. The monitor, or condensing unit, can be seen on the top of each appliance, looking like some kind of turret on a Civil War ironclad. At that time, many homes (especially those in rural locations) still used iceboxes that had to be supplied each week by the local iceman to keep foods cold. Even in the 1950s, the iceman could still be seen making such deliveries in many neighborhoods.

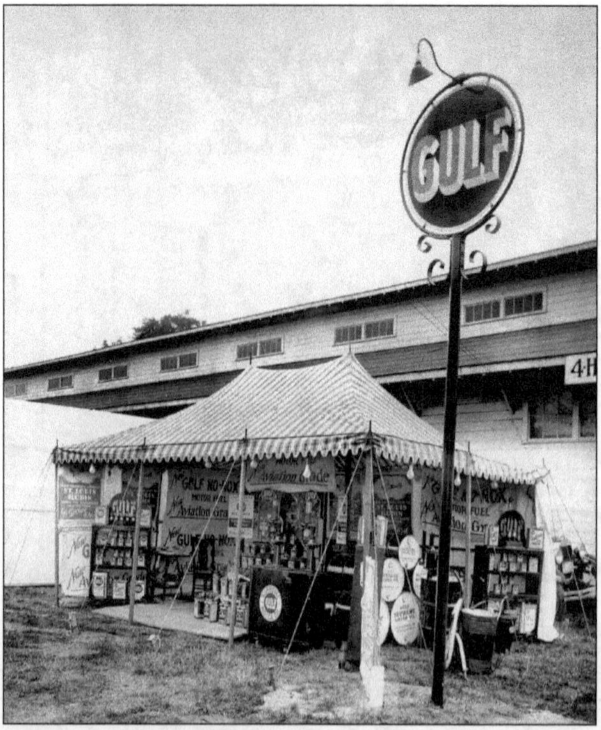

This very attractive commercial exhibit was located alongside the vegetable building in 1929. According to the sign inside the tent, the Gulf products included aviation-grade gasoline. Perhaps there was a connection to the nearby fledgling Beverly airport at this time. Also, a flying poster is hanging on the back wall, advertising the St. Louis-Robin endurance records.

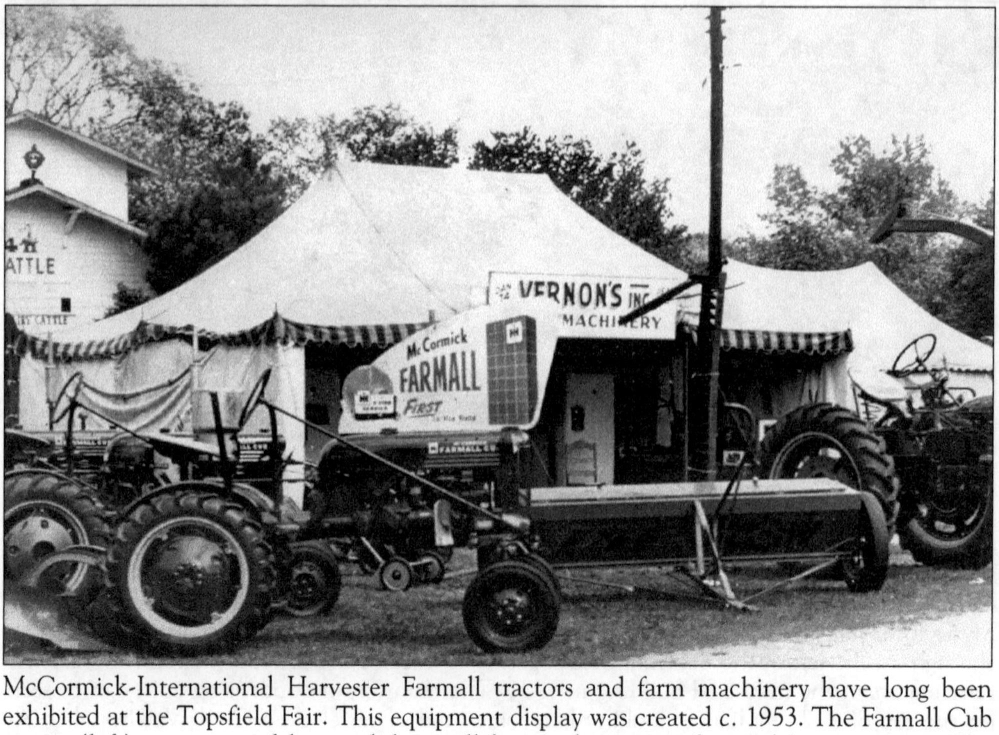

McCormick-International Harvester Farmall tractors and farm machinery have long been exhibited at the Topsfield Fair. This equipment display was created *c.* 1953. The Farmall Cub tractor (left) was very useful around the small farm and was manufactured for many years. The truck in this book's 1924 cover picture shows it to be advertising McCormick-International farm power.

Four

4-H Activities

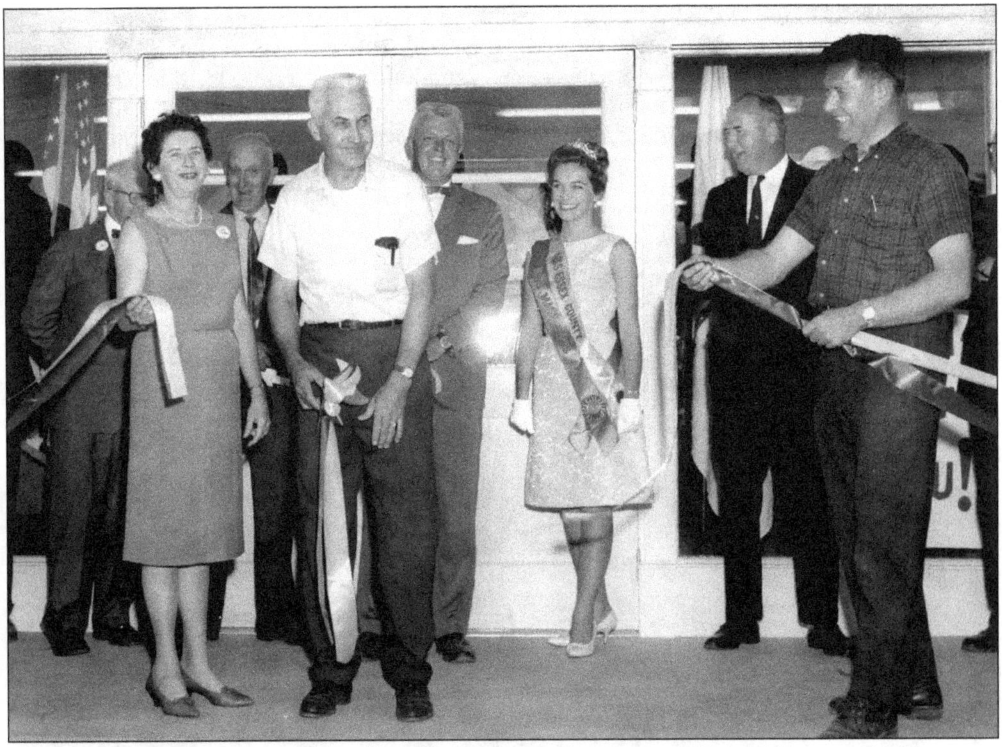

The ribbon-cutting ceremony officially opening the new 4-H building in 1963 was a momentous occasion for 4-H in Essex County. The building was constructed for this use by the Massachusetts 4-H Foundation. From left to right are Clifford Elliot, 4-H Club agent Margaret Faulkner, fair secretary Ellery E. Metcalf, William Owen, Topsfield Fair president Augustus Means, Miss Essex County Elaine Brant, unidentified, 4-H Club agent Charles Blanchard, and fair general manager Paul Corson. The 4-H Club activities until 1963 were held in what is now the sportsmen's building, which was originally built as a dance hall in 1921. The sportsmen's show had been in a tent in the area of Coolidge Hall before 1963.

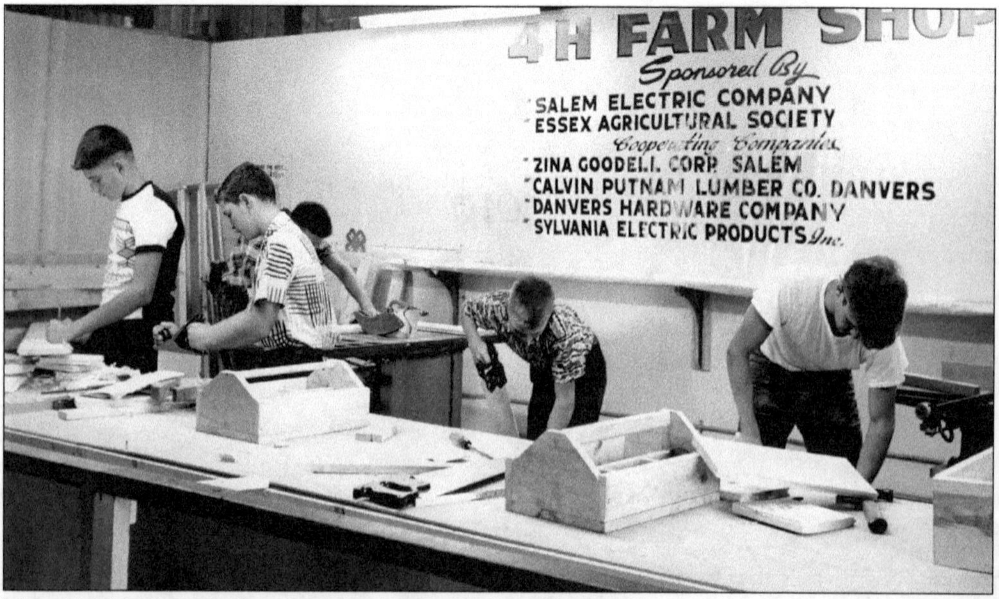

The 4-H building in the 1950s was the old dance hall, where the sportsmen's show is held today. This photograph shows a group of boys working on their woodworking projects in the 4-H Farm Shop, which was sponsored by the organizations listed on the rear wall. In 1963, the 4-H activities moved into their new building between the Grange and apiarian, or bee, building. (Photograph courtesy of Charles and Dorothy Blanchard.)

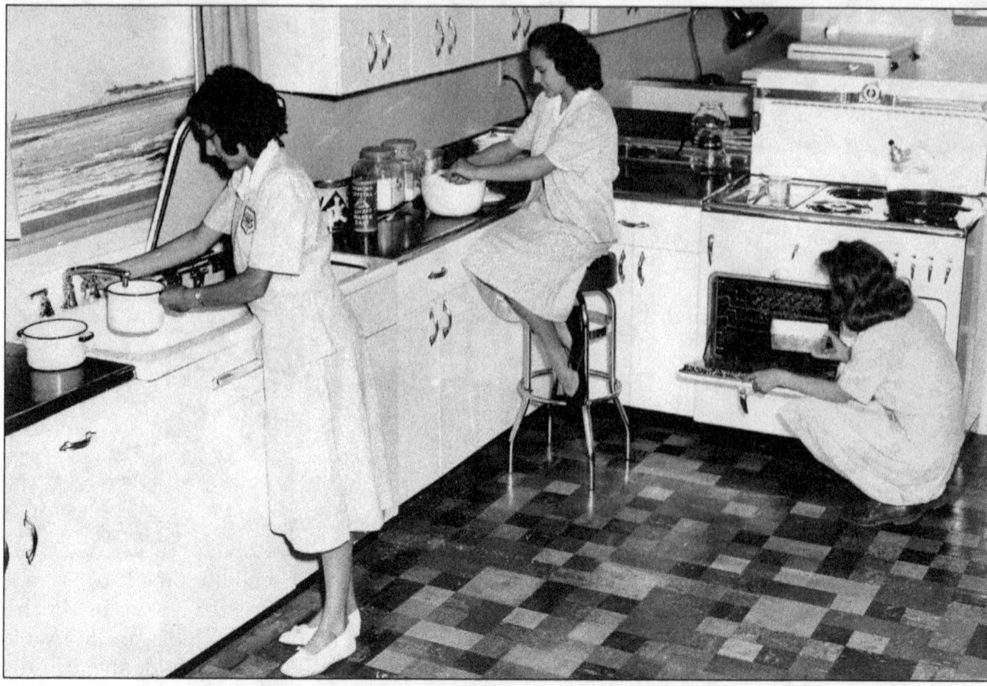

Girls work in their homemaking kitchen exhibit in the fair's 4-H building in the 1950s. The girls' exhibits were in the building on alternate years from the boys' shop exhibit (seen in the previous view). On several occasions, the girls prepared a special dinner for the governor of Massachusetts. (Photograph courtesy of Charles and Dorothy Blanchard.)

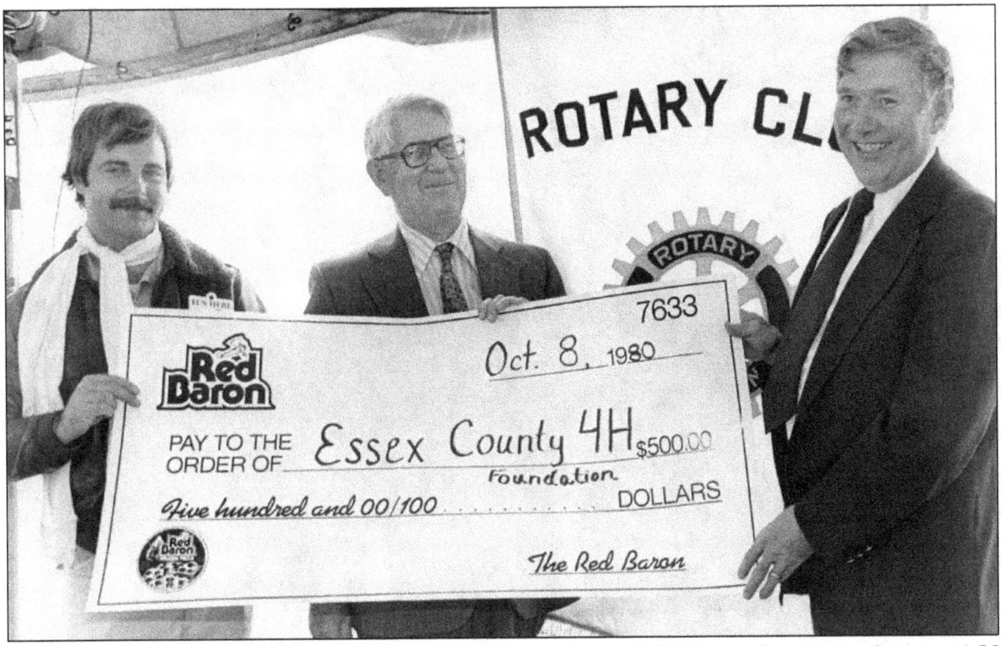

In October 1980, the Red Baron Pizza Company donated $500 to the Essex County 4-H Foundation in a ceremony at the fair. Holding the left end of the check is the "Red Baron" for that advertising aerial campaign over the Topsfield Fair, flying in his vintage biplane towing a large banner. In the center is William Spakowski, president of the Essex County 4-H Foundation. Completing the group is 4-H Club agent Charles Blanchard of Georgetown.

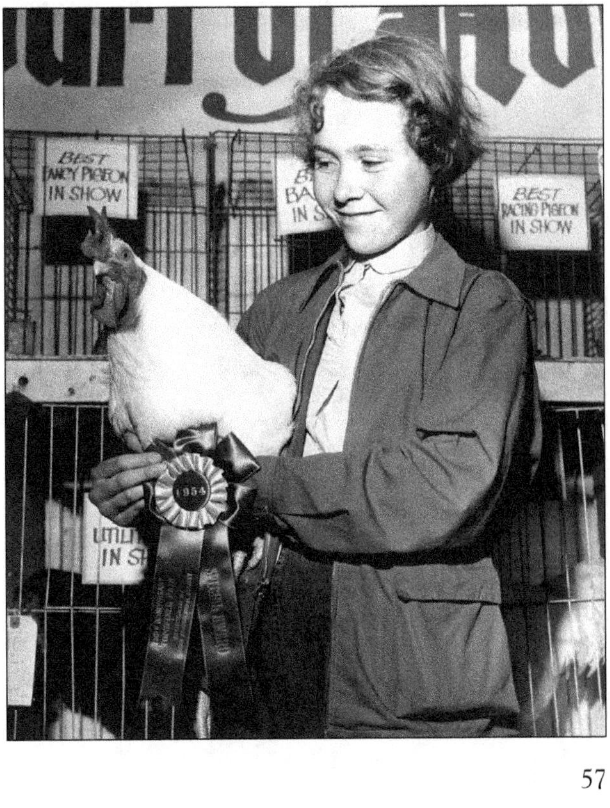

Ann Hall of Rowley shows off her White American cockerel in the 4-H poultry show in 1954. This cockerel, the winner of Best of Show in 4-H Poultry, will take his place in the "Court of Honor" cages behind Hall for the duration of the fair. This breed of poultry was developed in Saugus by Ellery E. Metcalf, the longtime secretary of the Essex Agricultural Society and instructor of vegetable gardening at the Essex Agricultural School in Hathorne. (Photograph courtesy of Charles and Dorothy Blanchard.)

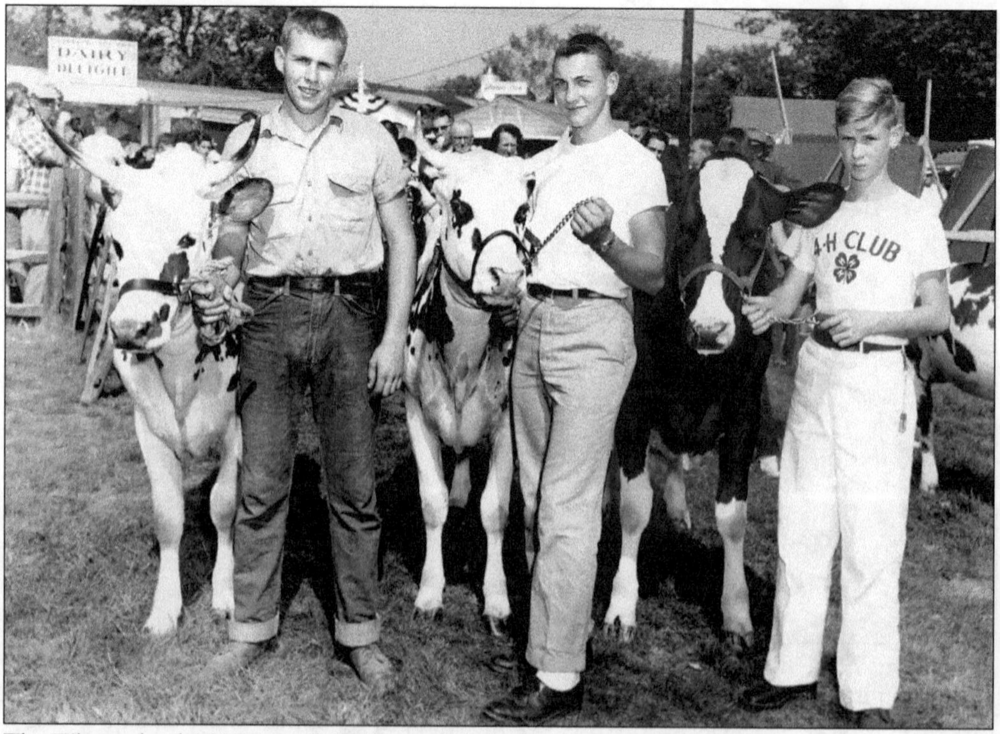

The Whittierland 4-H Dairy Club of Haverhill was well represented in the 4-H cattle competition at the 1955 fair. On the left is Leonard Kimball of Haverhill, with his Ayrshire heifer. In the center is John McCarthy of Haverhill with his Ayrshire heifer. On the right is John W. Riesen, also from Haverhill, with his Holstein heifer. John Riesen is now a research professor of animal science at the University of Connecticut. (Photograph courtesy of Charles and Dorothy Blanchard.)

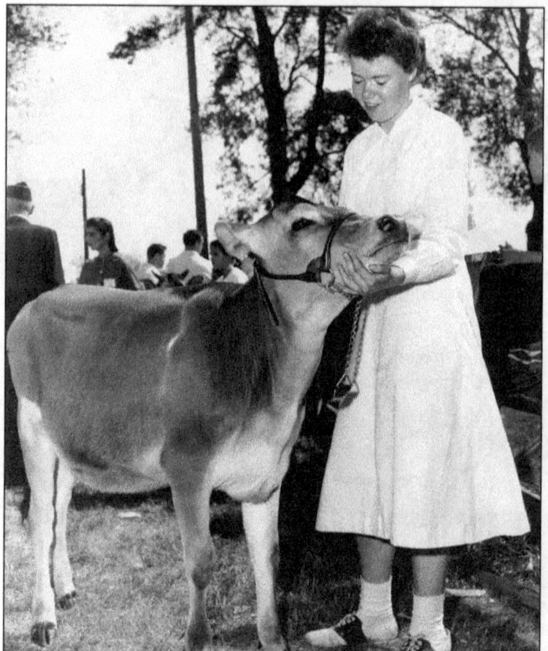

Lucy Hall of Groveland shows her Jersey heifer calf at the 4-H cattle show at the Topsfield Fair in the 1950s. (Photograph courtesy of Charles and Dorothy Blanchard.)

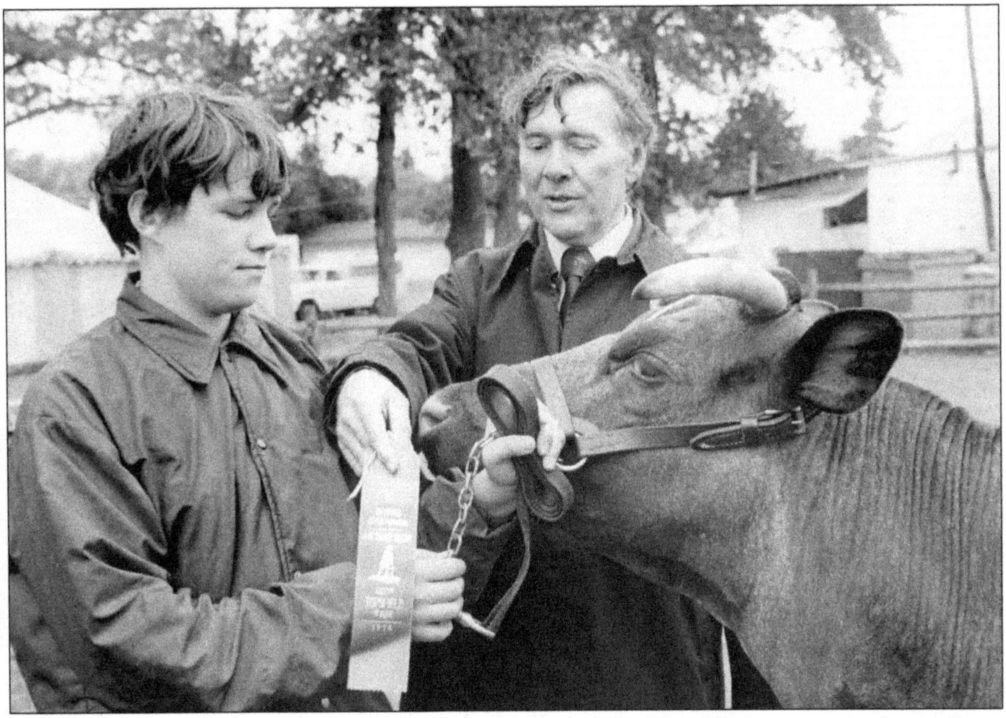

Billy Bartlett of Salisbury is presented with a championship ribbon for his prizewinning milking shorthorn cow in the 4-H cattle show by Essex County 4-H Club agent Charles Blanchard in 1976. (Photograph courtesy of Charles and Dorothy Blanchard.)

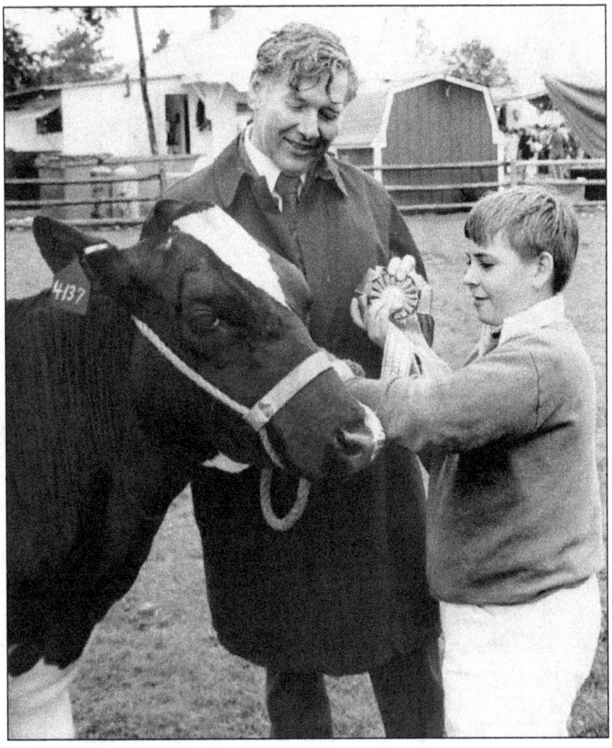

David Dunajski of Peabody is presented with a championship ribbon for his prizewinning milking Holstein cow in the 4-H cattle show by Charles Blanchard in 1976. (Photograph courtesy of Charles and Dorothy Blanchard.)

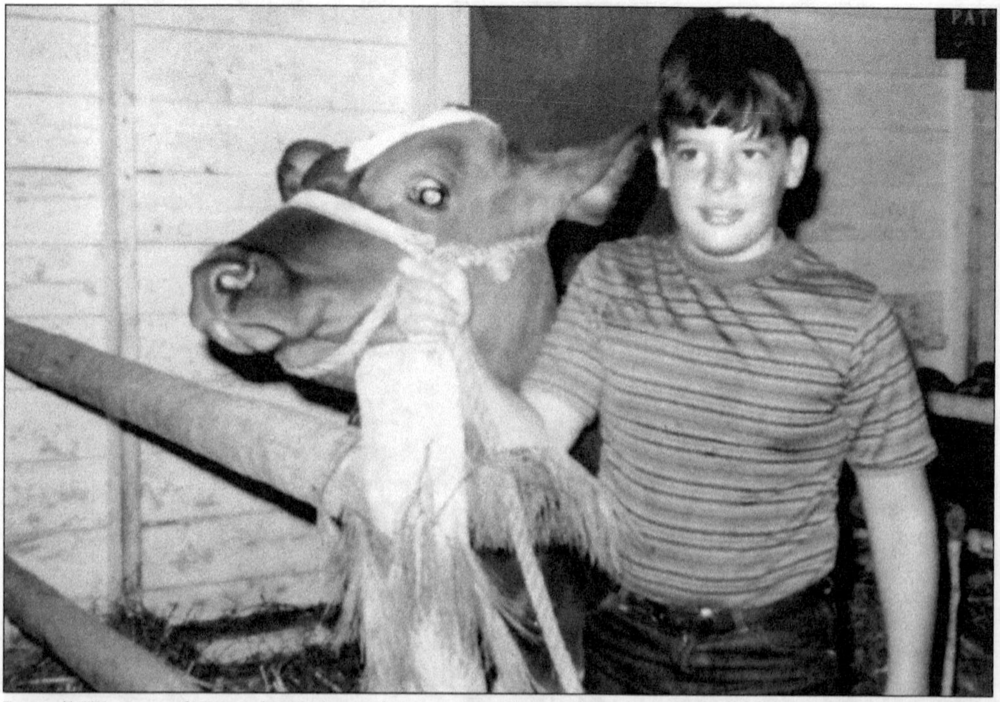

Russell Waring of Groveland shows his registered Ayrshire heifer, Mabel, at the Topsfield Fair in 1970. His father was the manager of the Merideth Farm in Topsfield, and Russell is now the head of the Topsfield Fair maintenance department.

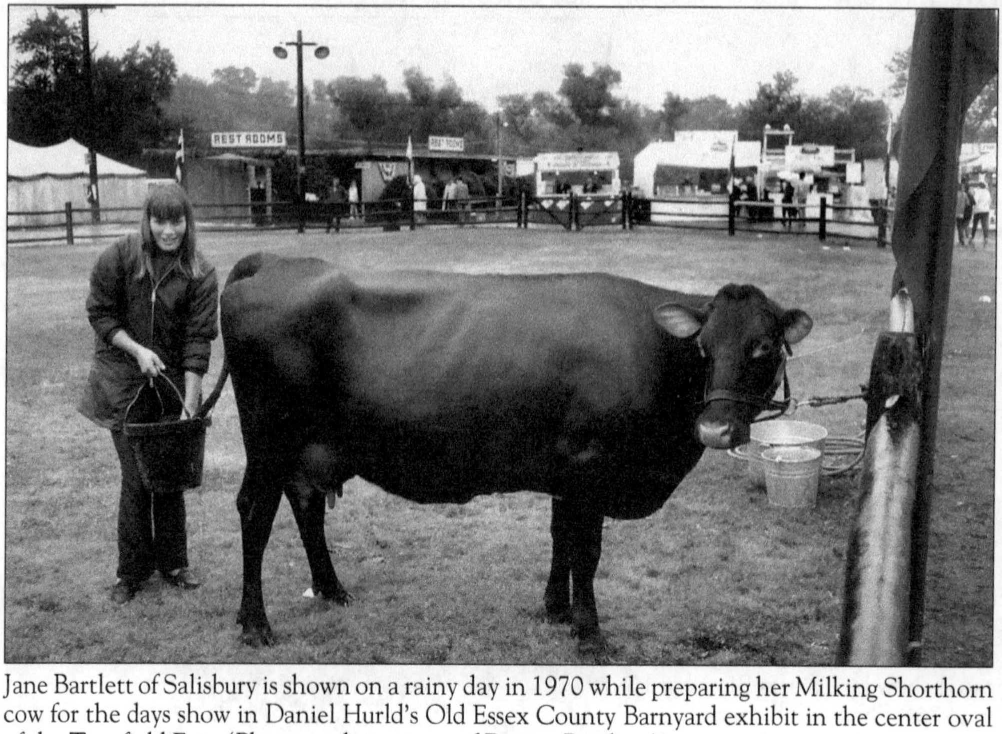

Jane Bartlett of Salisbury is shown on a rainy day in 1970 while preparing her Milking Shorthorn cow for the days show in Daniel Hurld's Old Essex County Barnyard exhibit in the center oval of the Topsfield Fair. (Photograph courtesy of Donna Bartlett.)

Elizabeth Smith from Groveland wins the 4-H Sheep Special Award with her Shropshire ewe in 1955 at the Topsfield Fair. (Photograph courtesy of Charles and Dorothy Blanchard.)

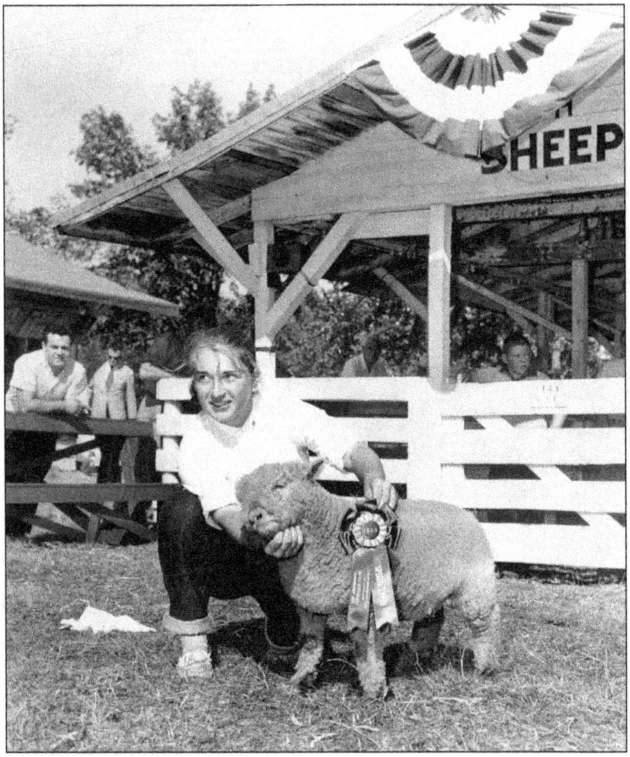

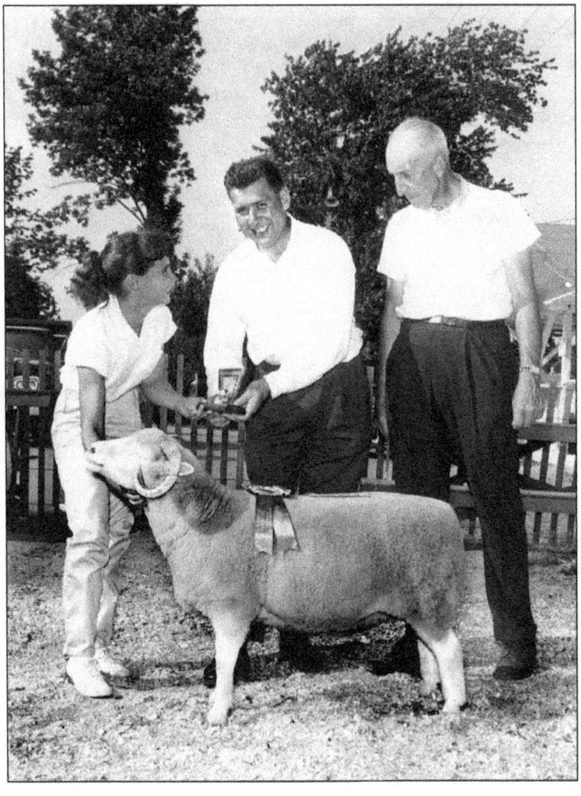

In 1959, Judith Owen, daughter of 4-H board member Bill Owen, wins the 4-H sheep show championship with her Dorset ewe. The award is being presented by 4-H Club agent Charles Blanchard. The judge standing next to Blanchard is Loren "Sy" Tirrell, a professor of animal science at the University of New Hampshire. (Photograph courtesy of Charles and Dorothy Blanchard.)

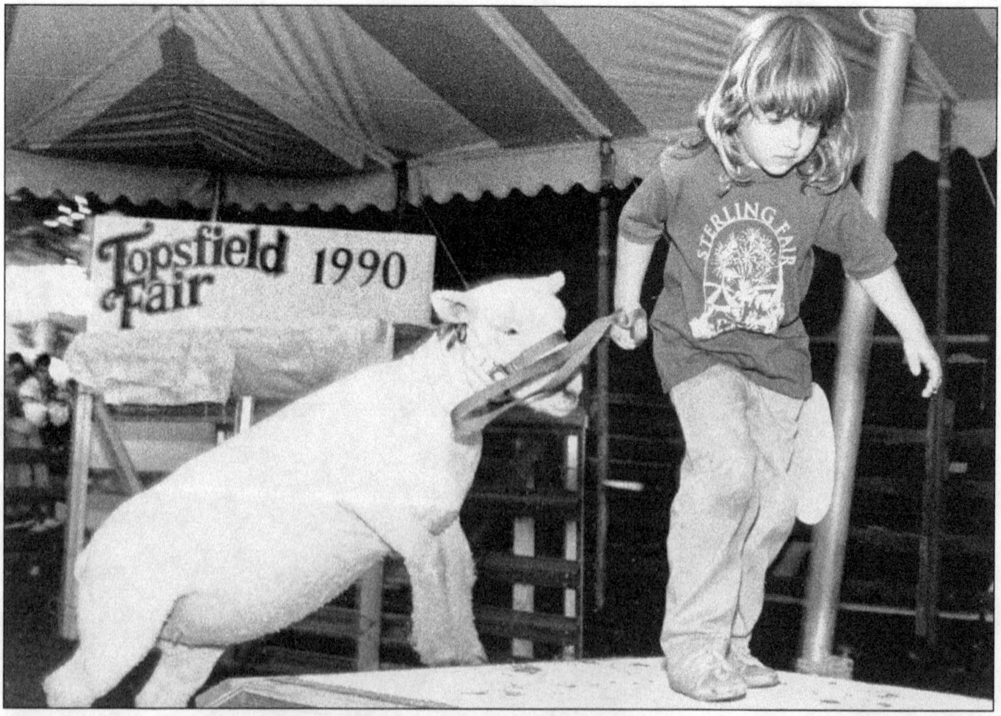

Alicia Buffington of Rehoboth helps her reluctant sheep over the bridge in the obstacle course during the 1999 junior competition at the fair.

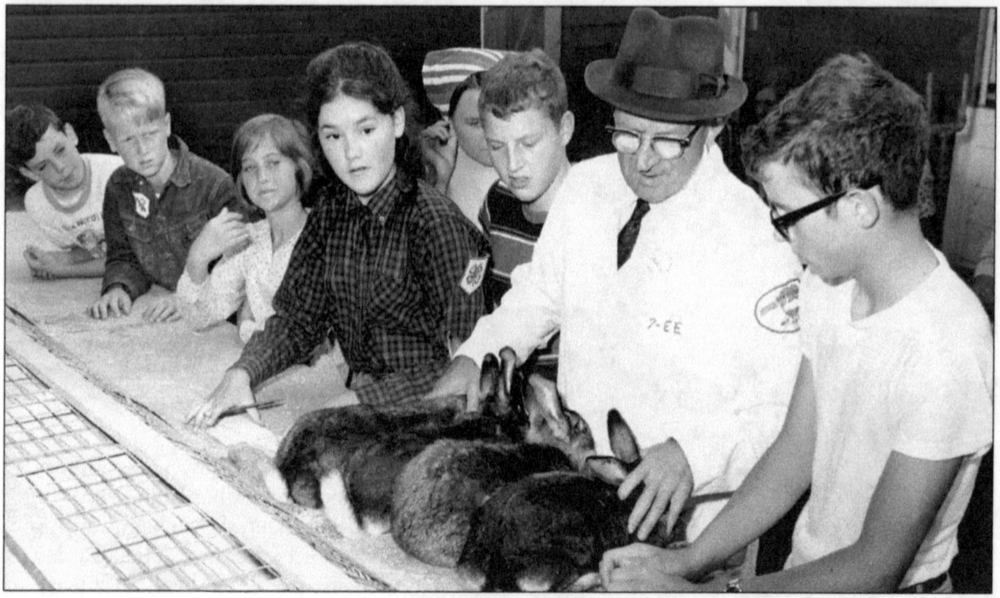
Members of leader Ormund Scott's Great Scott Rabbit 4-H Club, in Middleton, look on intently as their rabbits are being judged in the rabbit and cavie building at the 1954 fair. On the far right is George Roaf of Danvers, soon to become a student at the Essex Agricultural School in Hathorne. Roaf's parents, George and Virginia Roaf, have been longtime committee members of the flower shows. Virginia has served as the show's clerk and secretary for many years, and George has served on the judging panel many times. (Photograph courtesy of Charles and Dorothy Blanchard.)

This 4-H rabbit club member, nine-year-old Richard Coglano of Canton, was one of the young competitors in the 4-H rabbit competition at the fair.

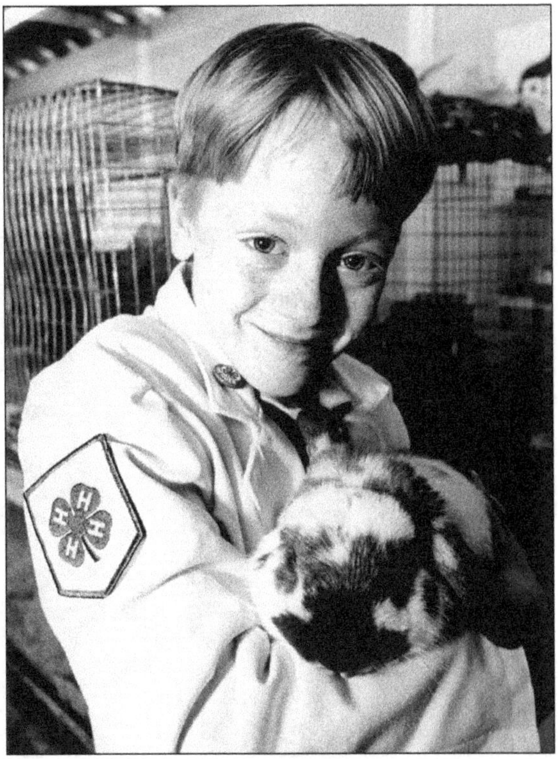

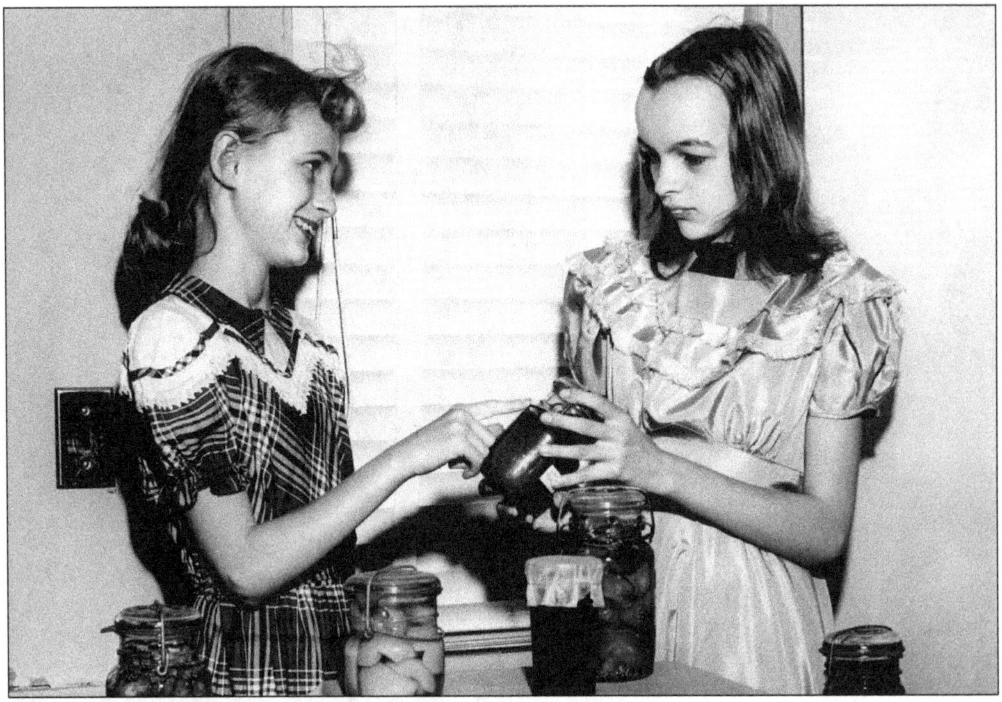

Discussing their canning project for the upcoming 4-H competition at the 1953 fair are 11-year-old Ivy Roberts and 14-year-old Henrietta Colby, both of whom were class winners the year before. Both are from Saugus. Their sisters are seen on the next page.

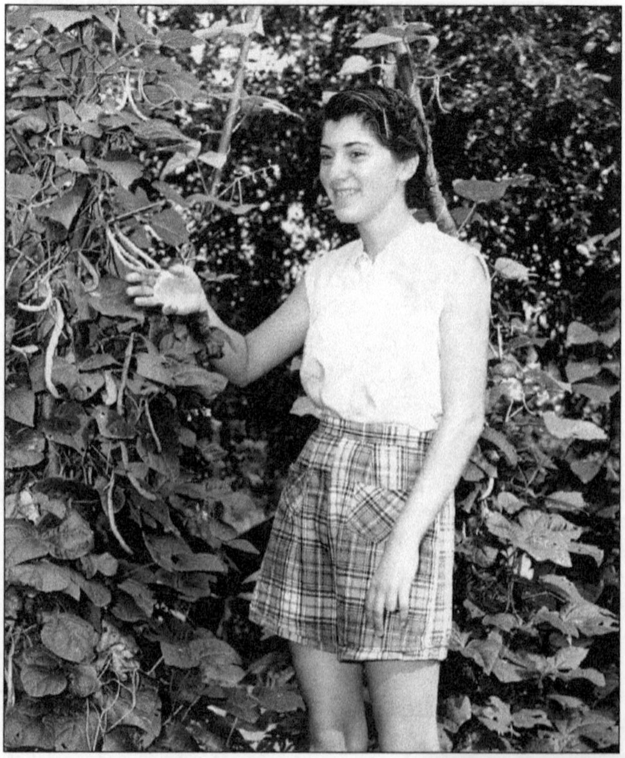

Looking at her crop of beans that she plans to enter at the 1953 Topsfield Fair is Eleanor Metcalf of Saugus. Metcalf was a major winner in the 4-H show the previous year. She is the daughter of Ellery E. Metcalf, a longtime fair secretary and chairman of the vegetable show committee. Ellery also taught vegetable garden culture at the Essex Agricultural School.

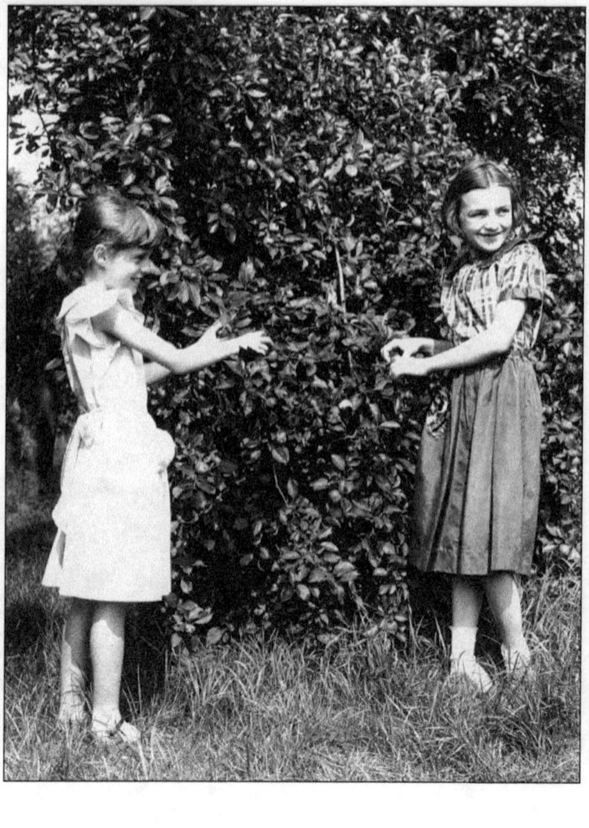

For their 4-H exhibit at the 1953 Topsfield Fair, Alexandra Roberts and Lorette Colby, both of Saugus, look over apple trees from which they will pick fruit for displaying.

Five

Special Awards

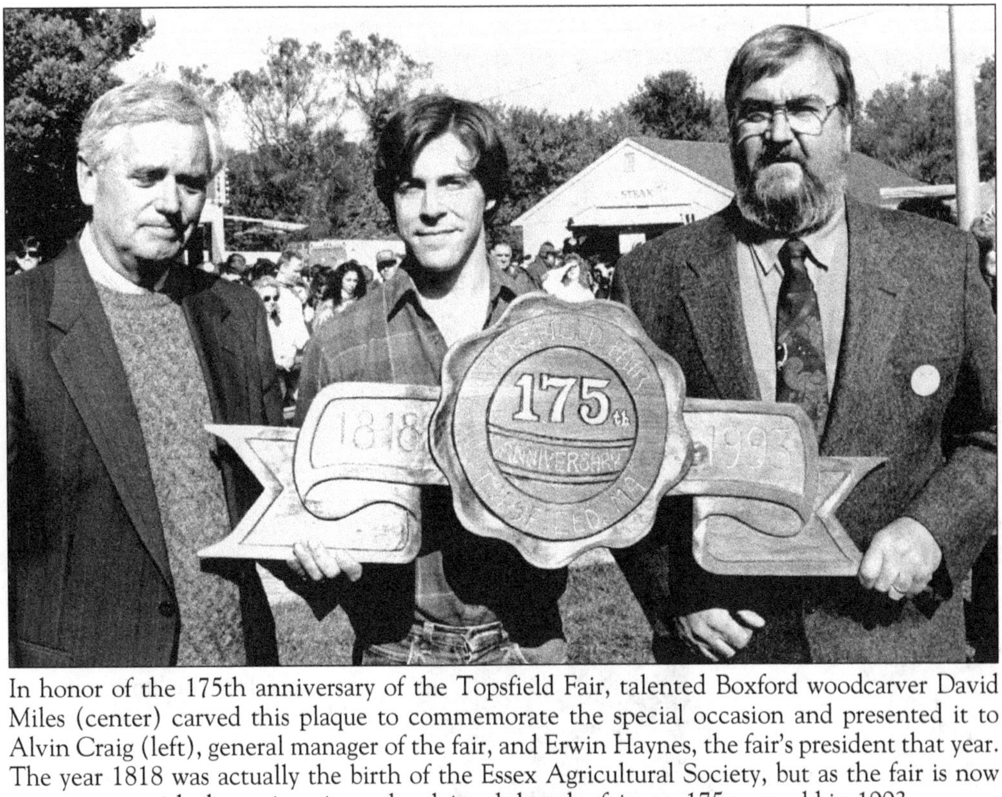

In honor of the 175th anniversary of the Topsfield Fair, talented Boxford woodcarver David Miles (center) carved this plaque to commemorate the special occasion and presented it to Alvin Craig (left), general manager of the fair, and Erwin Haynes, the fair's president that year. The year 1818 was actually the birth of the Essex Agricultural Society, but as the fair is now synonymous with the society, it can be claimed that the fair was 175 years old in 1993.

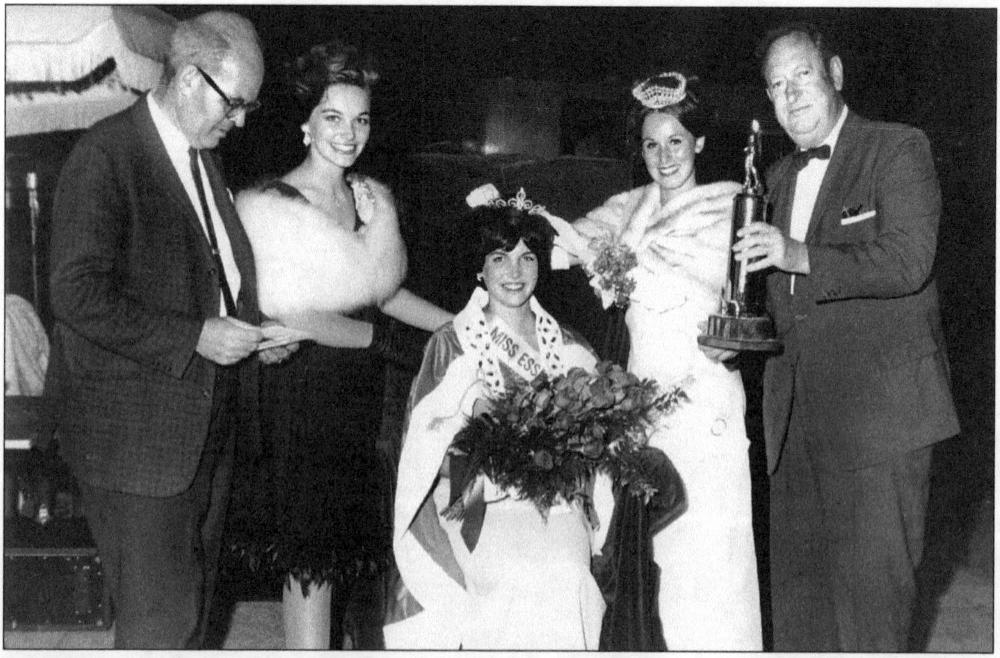

Paul Corson (left), general manager of the Topsfield Fair for many decades, assists Warren Rockwell, the fair's president from 1964 to 1965, with the ceremonies. Next to Paul Corson is Elaine Brant, a former Miss Essex County and chairman of the Miss Essex County Pageant committee. The lovely new queen is being crowned by the reigning queen.

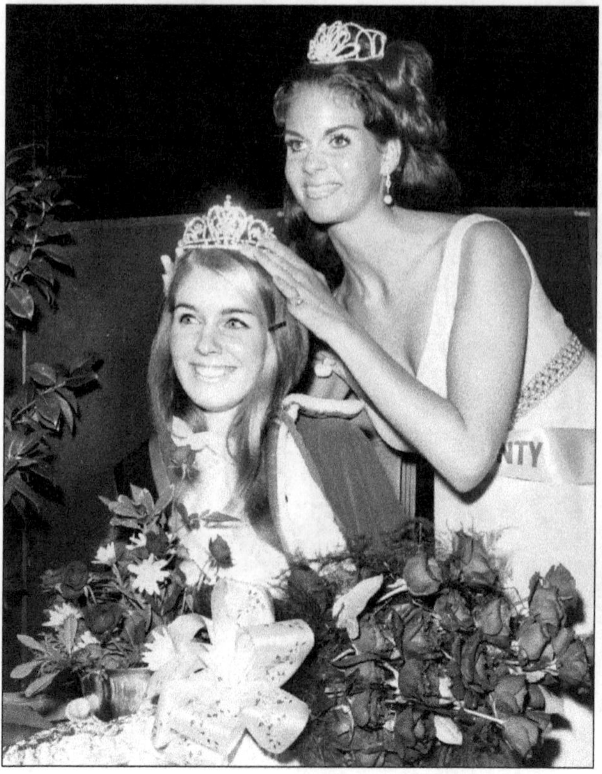

Sharon Freeman of North Andover was named Miss Essex County for the 1970 season at the Topsfield Fair in 1969. Freeman was going off to the University of Michigan to become a music teacher. She is being crowned by the reigning 1969 Miss Essex County, Phyllis Hossfelt.

For years, the fair has been host to many types of pageants. There have been longstanding Miss Fair, Miss Essex County, and (most recently) Mrs. Essex County contests in which winners are crowned and will reign for one year. Cathy Coleman of Lynn, shown in front of the old fair administration building, was crowned Miss Teen World and attended many fair events.

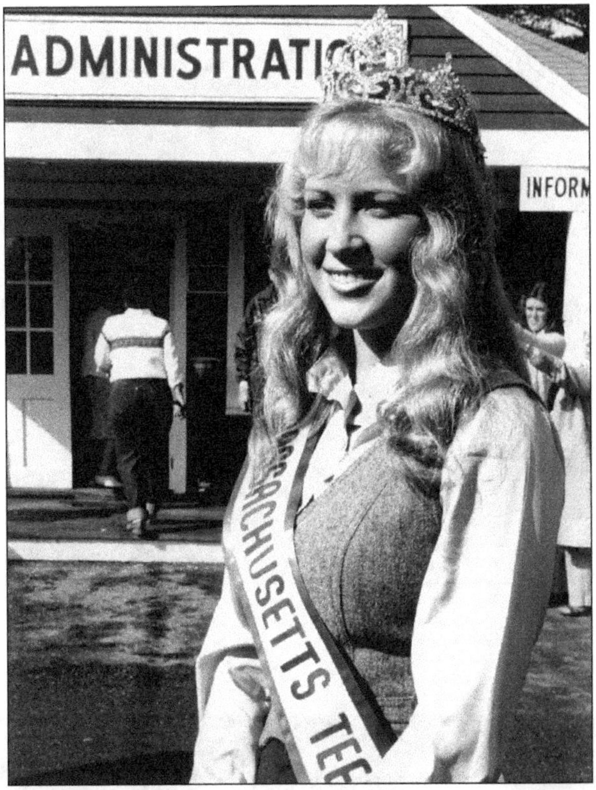

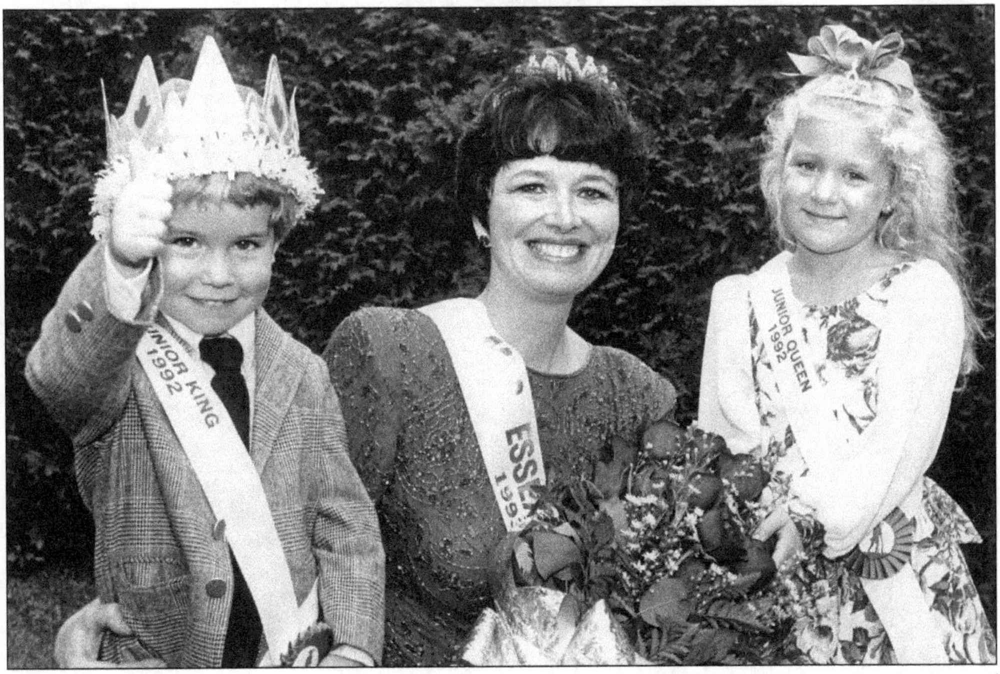

The 1992 celebrities include Caleb Nobel of Rowley as Junior King, Marion Garabedian of Peabody as Junior Queen, and the 1993 Mrs. Essex County, Brooke Halinen of Danvers.

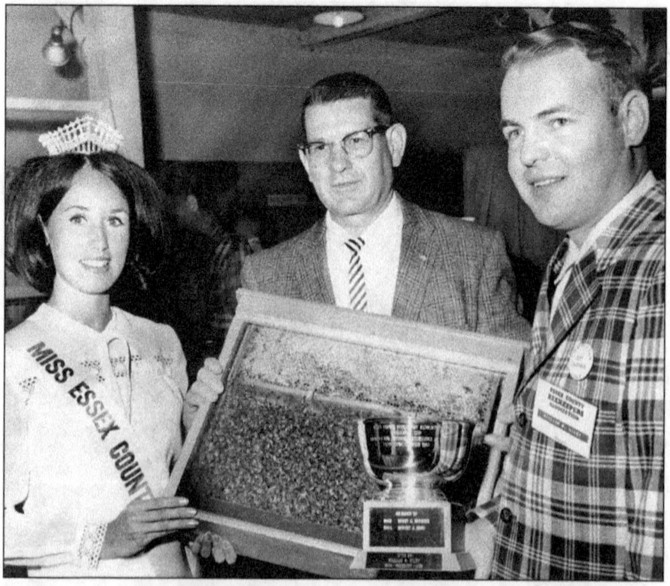

The 1965 Miss Essex County helps William Wiley of Wenham present the Essex County Beekeepers Challenge trophy to Henry Neunzer of Andover for exhibition excellence in the apiarian department at the fair.

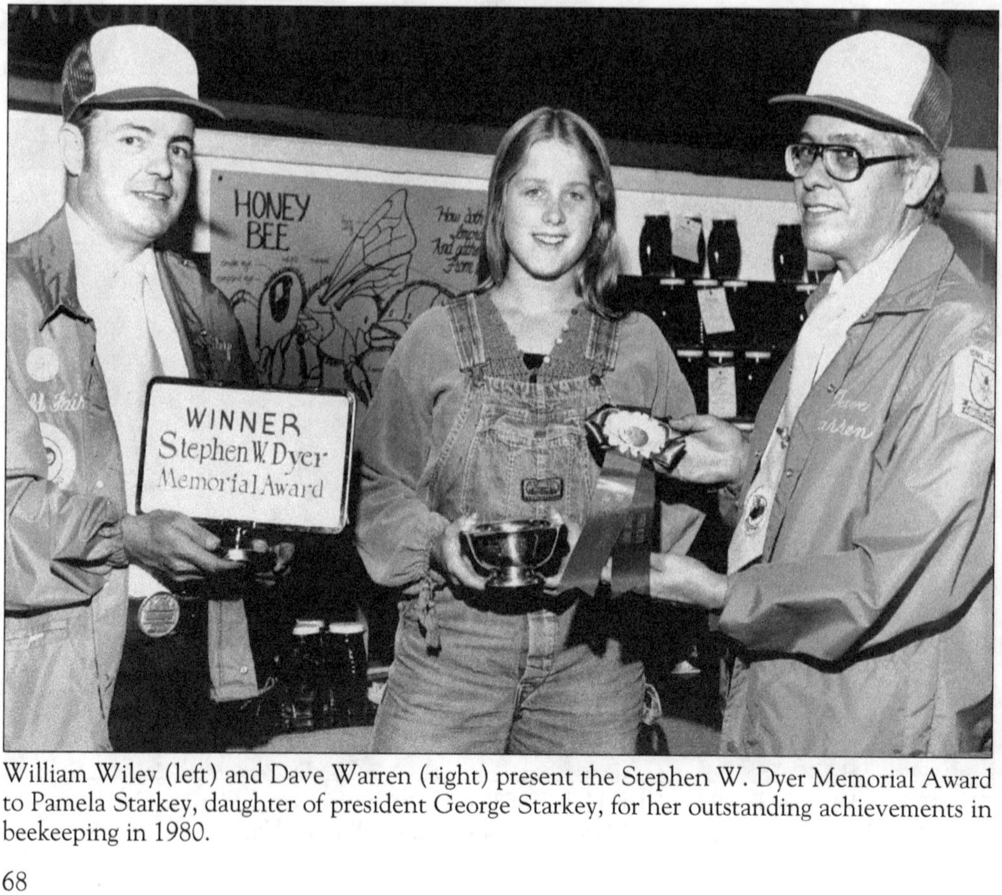

William Wiley (left) and Dave Warren (right) present the Stephen W. Dyer Memorial Award to Pamela Starkey, daughter of president George Starkey, for her outstanding achievements in beekeeping in 1980.

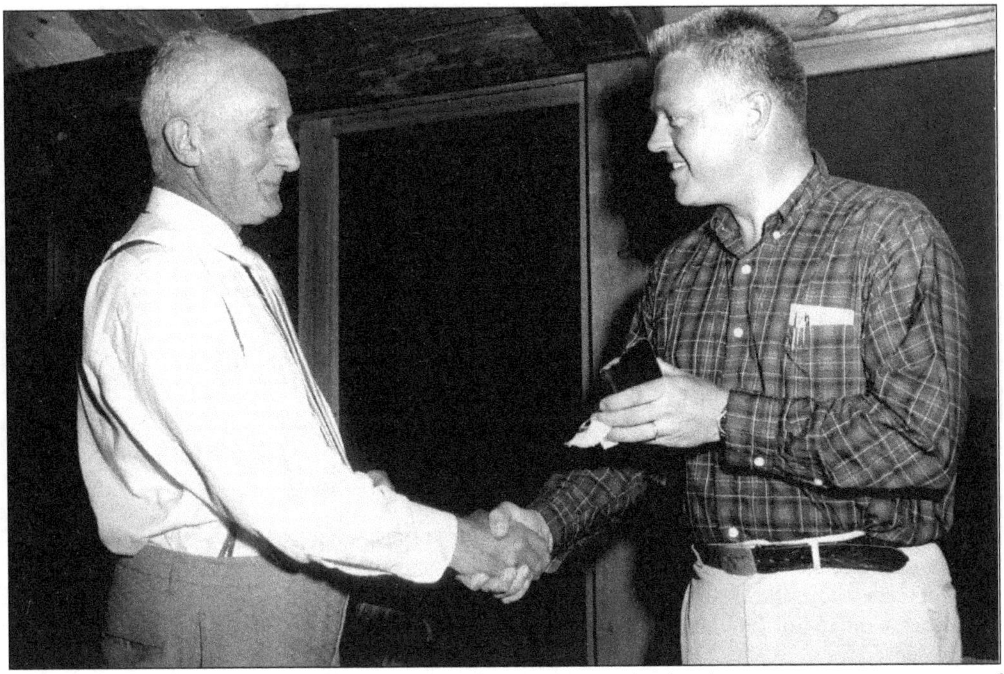

Ellery E. Metcalf of Saugus receives special recognition for his long service as secretary of the Essex Agricultural Society, the parent organization that has put on the fair for 185 years. Ellery was a well-respected and loved gentleman who taught vegetable gardening at the Aggie. For years, he was the chief FFA advisor there. He taught the author vegetable gardening and groomed him for a 20-year stint as chief advisor to the Essex Chapter FFA after Ellery retired c. 1964. Ellery was also the developer of the White American breed of egg-producing poultry, some of which were even shipped to the pope in the Vatican.

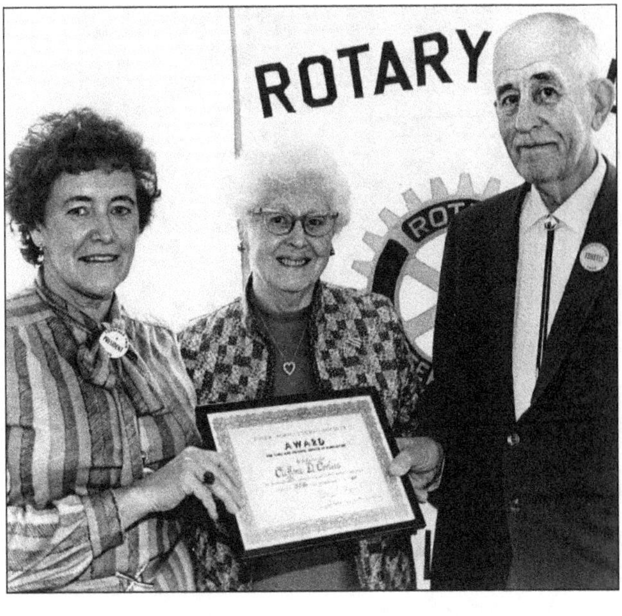

"Fifty five Years of Service to the Fair" reads this certificate presented to Mr. and Mrs. Clifford D. Corliss of Hamilton by Nancy E. Begin. The award was given during one of the fair's ceremonies honoring the many devoted associates working in the agricultural and other fair departments. The Corlisses were the owners and operators of the Corliss Brothers Nursery in Ipswich. Many fine garden exhibits were put into the fair's flower show over the years, and the Corliss Nursery and Garden Center had continued to exhibit at the flower show for many years, although under new ownership.

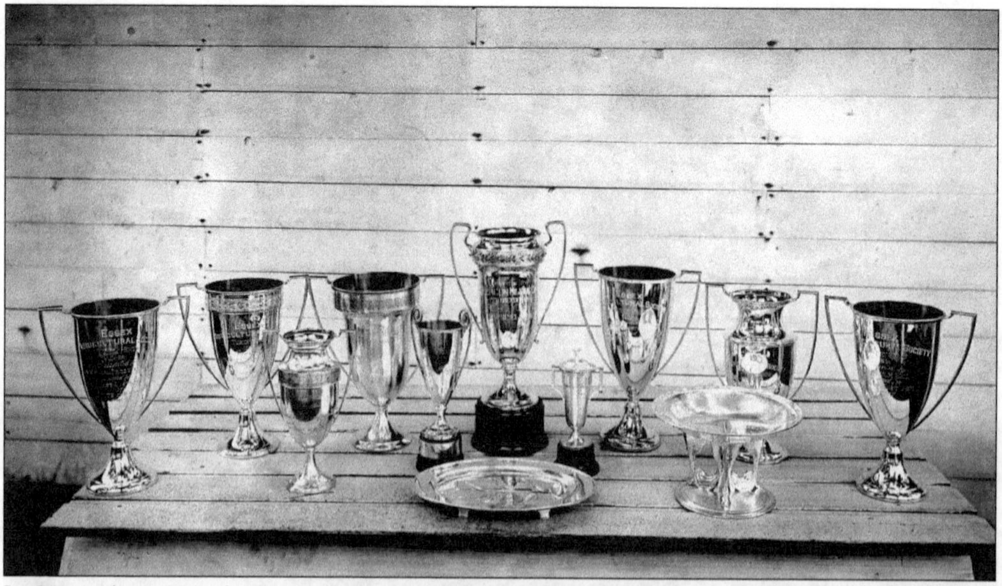

In September 1929, the Essex Agricultural Society awarded these beautiful loving cup trophies to various winners in the Turkey Show. Each is inscribed with the class in the turkey show it was presented for.

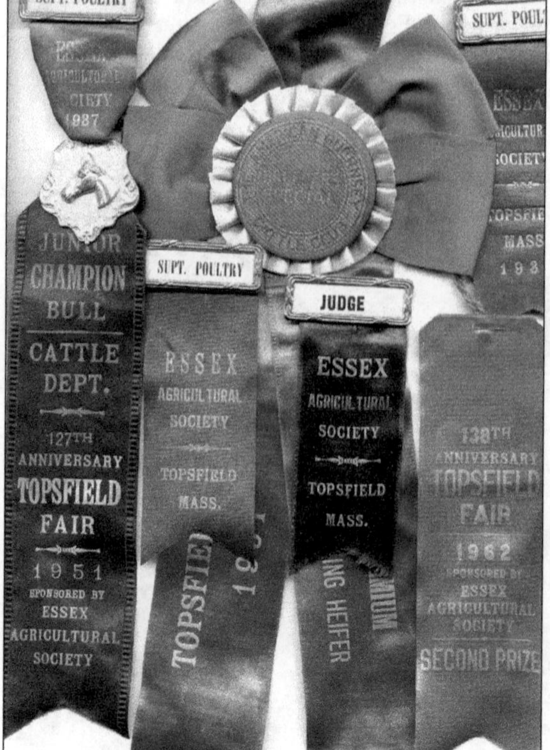

Here is a sampling of the various styles of ribbons given out as premiums for poultry and cattle classes from the 1920s to the 1950s.

Seven

Animals and Demonstrations

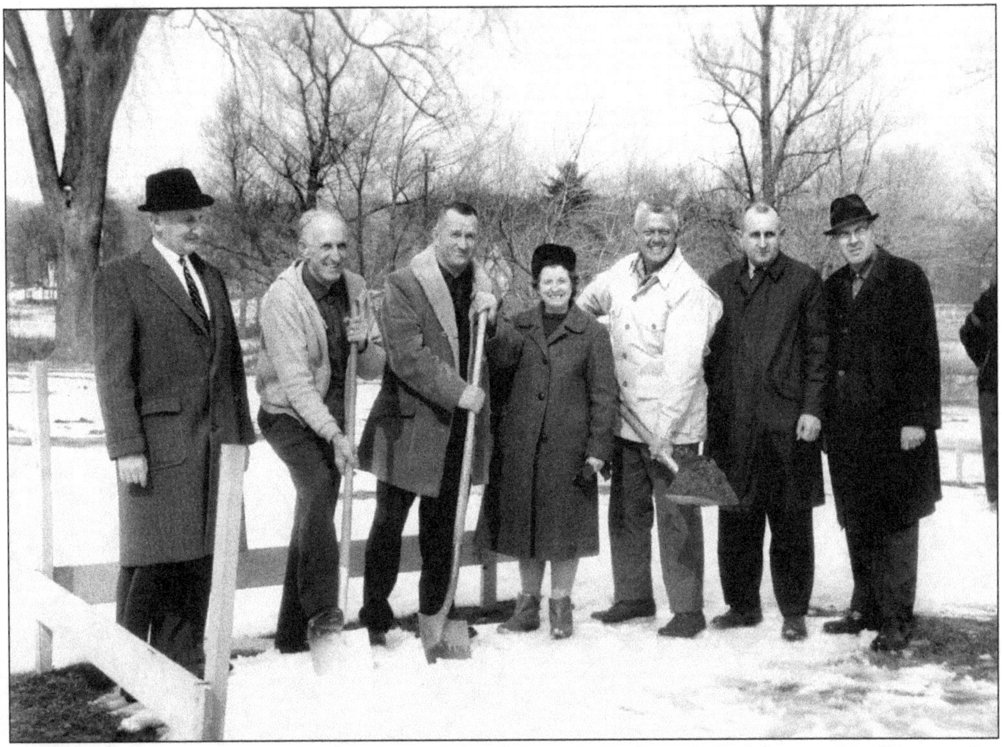

Here we see the groundbreaking ceremony for the new metal cattle building and milking parlor on March 14, 1964. Participating in the ceremony are, from left to right, fair secretary Ellery E. Metcalf of Saugus, also a vegetable gardening teacher at the Aggie; John P. Tutko, an executive board member and past fair president from Danvers; executive board member Harold T. Fletcher of Topsfield, the author's uncle; board member Beatrice Ropes; Augustus Means of Ipswich, the fair president at the time; Henry J. Ranta of Ipswich, board member and teacher of dairy science at the Aggie; and Paul Corson of Saugus, the fair's general manager from 1948 to 1985. In 1985, Paul's son Mark was his assistant manager, and then Mark managed the fair for the year 1986 until Alvin Craig of Essex was hired to assume the position of general manager.

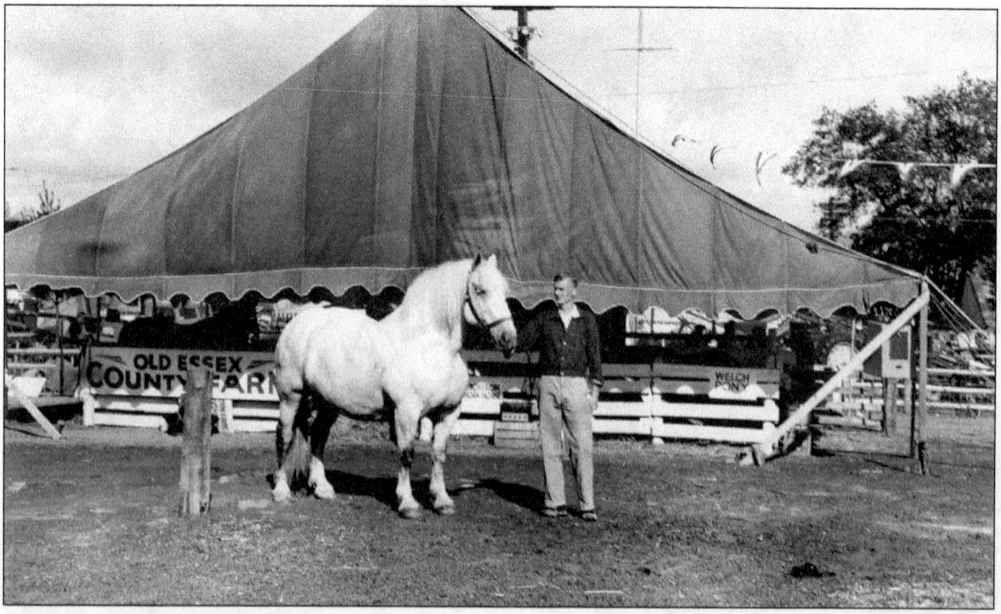

Daniel Hurld holds this magnificent gray Percheron draft horse mare at his Old Essex County Farm exhibit in the center oval of the fair in 1953. The 2,000-pound horse was from Appleton Farms, owned by Col. Francis Appleton, in Ipswich. In his first year of hosting the agricultural exhibitions at the front of the fair, Hurld featured the horse all week long in the barnyard. (Photograph courtesy of Barbara and Daniel Hurld.)

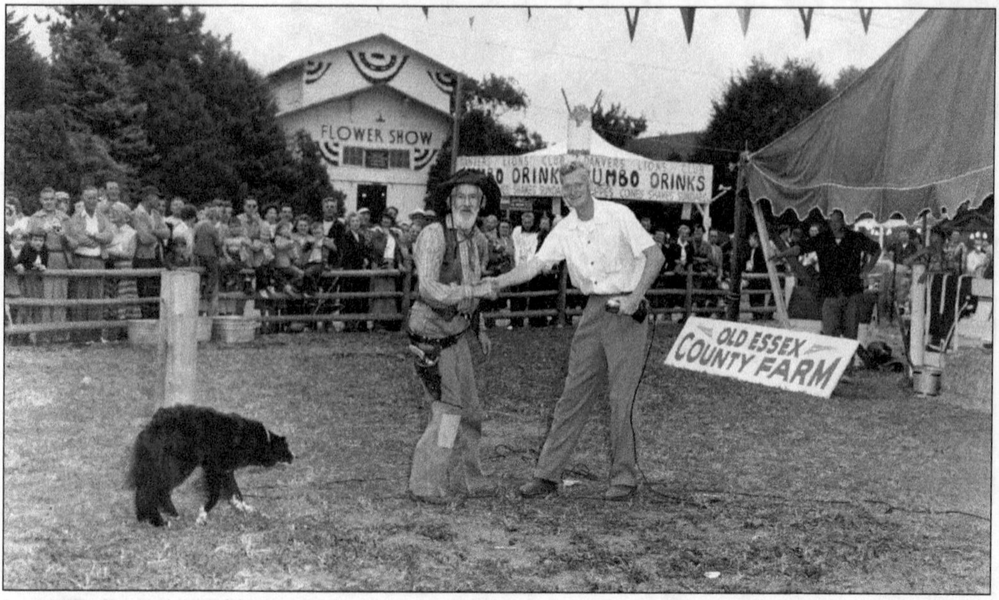

In 1955, Daniel Hurld greets the famed Western movie star Gabby Hayes to the front fair oval and the Old Essex County Farm, where Hurld promoted agriculture and presented various demonstrations, mostly involving agricultural practices to educate the public about the more rural way of life. Daniel's border collie, Bonnie Lass, acts more than suspicious toward Gabby; perhaps it is his unfamiliar appearing dress and sidearm that alarms her. (Photograph courtesy of Barbara and Daniel Hurld.)

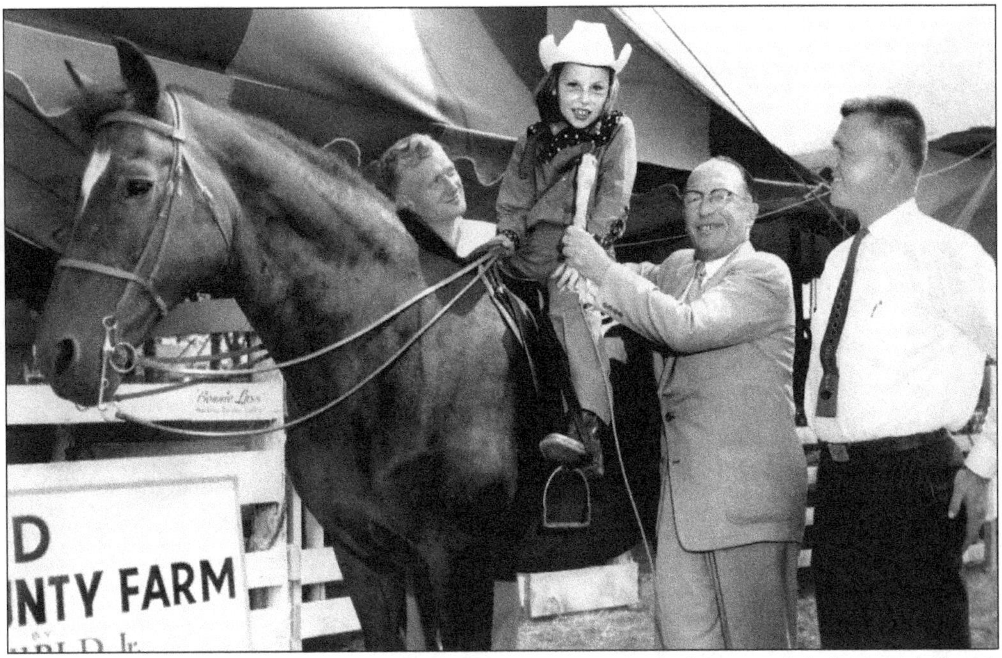

Nancy Hurld sits astride her mother Barbara's Morgan mare as her proud dad, Daniel Hurld, who hosts the Old Essex County Barnyard exhibit each year, looks on from behind the horse as the announcer asks her questions. Nancy has not been riding this mare, as she is in western attire and the stirrups on the English saddle are way below her boots. Augustus Means, president of the Topsfield Fair from 1956 to 1963, looks on.

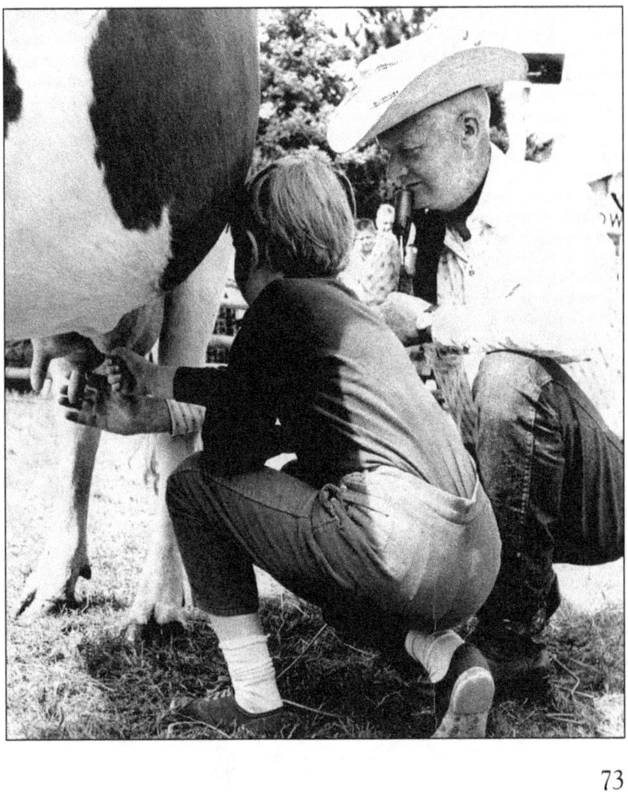

Elizabeth Nason, daughter of Ruth and Henry Nason of West Boxford, tries her hand at milking a cow under the instruction of Daniel Hurld in his barnyard exhibit at the Topsfield Fair in 1965. The Nasons were well known for their years of serving barbequed chicken to fair patrons from their concession stand near the center oval. (Photograph courtesy of Ruth Nason.)

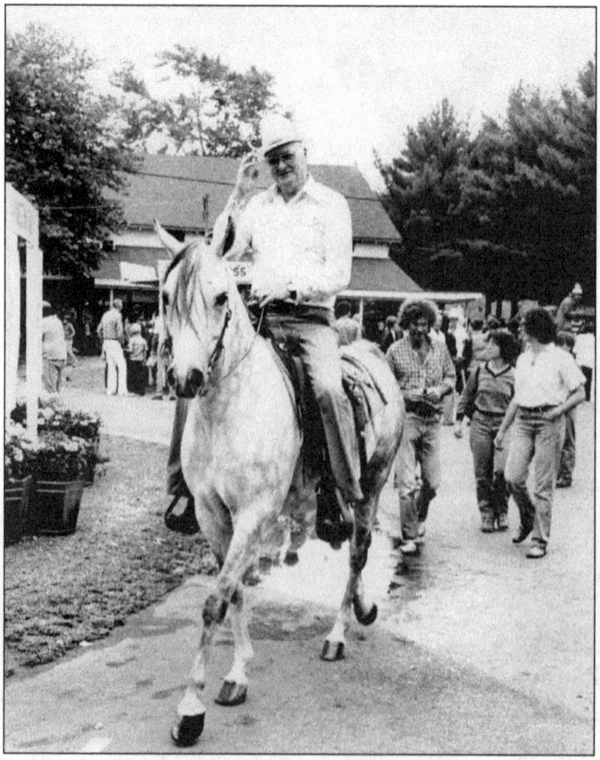

Daniel Hurld has just finished participating in the fair's 1979 opening-day parade to the fairgrounds from the center of the town of Topsfield. Hurld is riding his favorite Arabian mare, Kismet, which he proudly exhibits in his Old Essex County Barnyard in the center oval. (Photograph courtesy of Barbara and Daniel Hurld.)

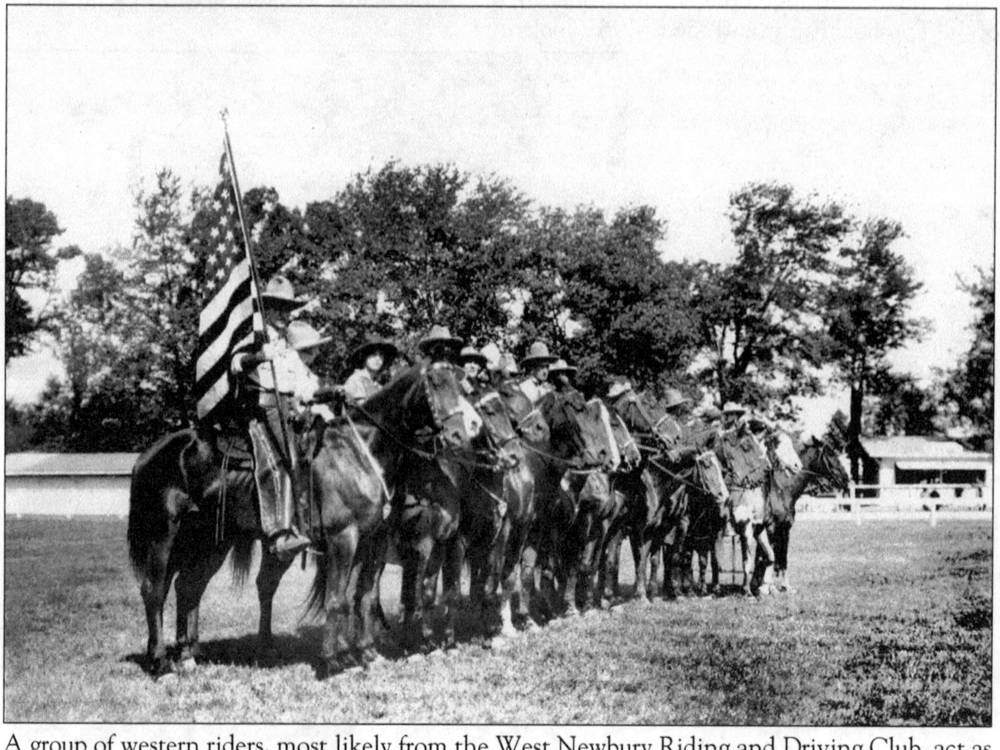

A group of western riders, most likely from the West Newbury Riding and Driving Club, act as color guard at a horse show in 1926 at the Essex County Fair in Topsfield.

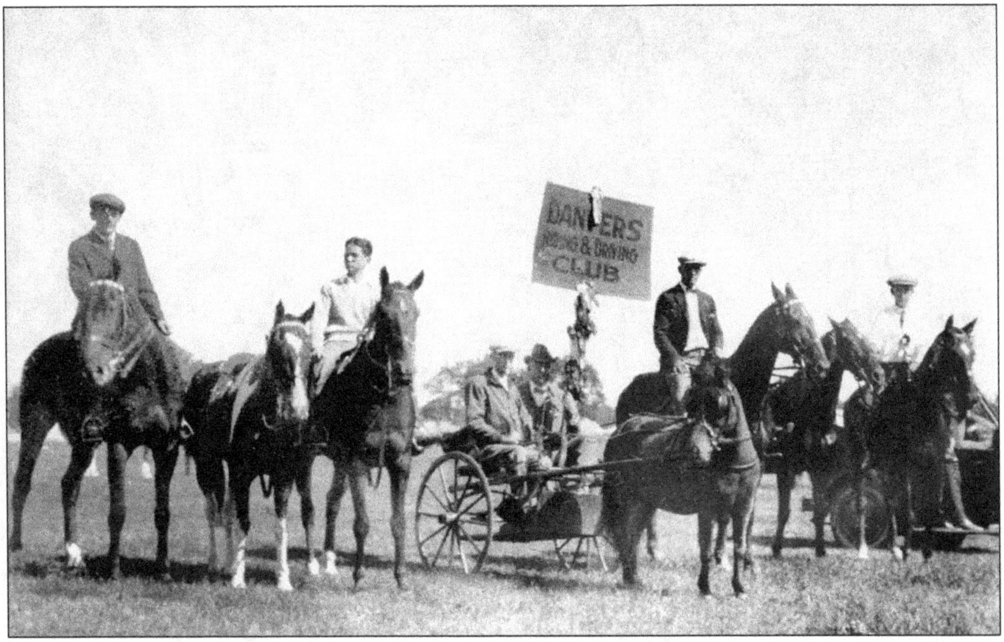

Several members of the Danvers Riding and Driving Club have their photograph taken at the fair. This club was active in the area for many years. Members were very diverse in their activities, as we see some members driving and others are riders attired in the English style.

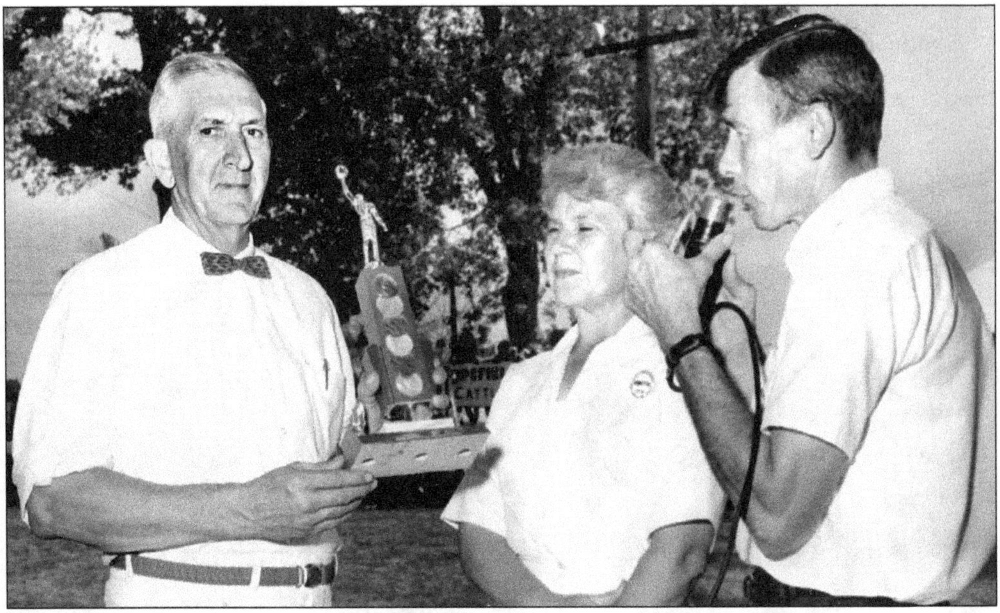

In 1968, Earl Foster (shown here with his wife, Beatrice) was presented with the Outstanding Farm Award in the 1950s. Rocky Knoll Farm, in North Andover, is where the Fosters housed their prizewinning Ayrshire dairy herd. The Fosters were longtime exhibitors at the Topsfield Fair. Some members of the family showed cattle in the 1800s, when the fair was called the Cattle Show and Exhibition, and after that at the Essex County Fair. The award is being presented by Charles DeGrandphree of Topsfield.

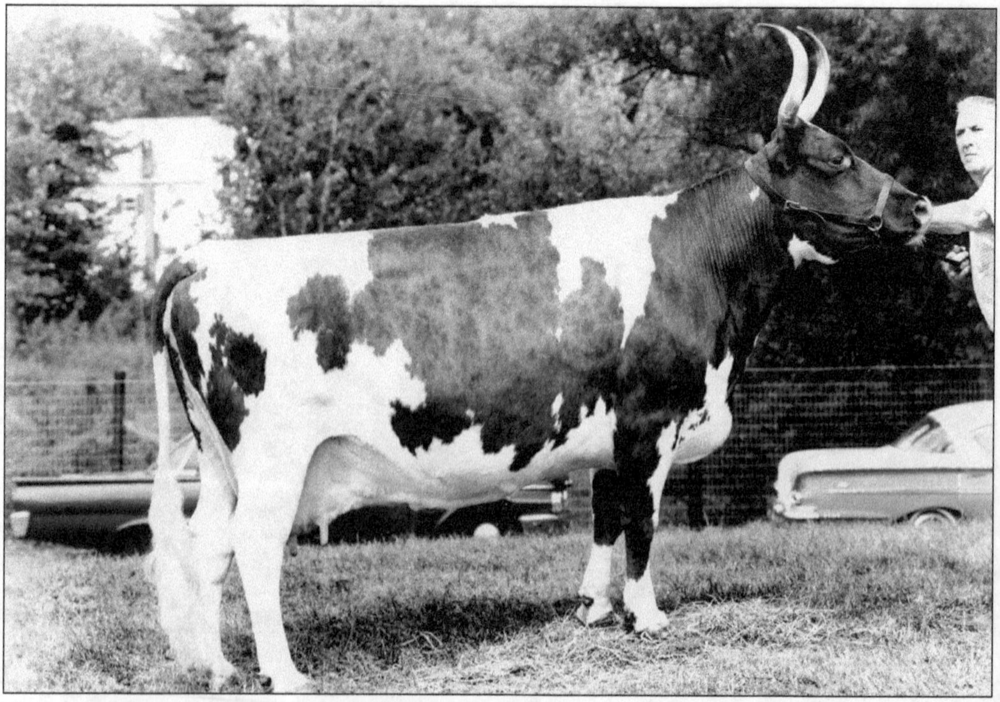

This great photograph, taken in 1959, shows a prizewinning Ayrshire cow "in hand" by Earl Foster of Rocky Knoll Farm in North Andover. Until the 1970s, most breeds of cattle had natural horns, but due to safety for both humans and other cattle, most all cattle are now dehorned soon after birth. There are a few breeds of cattle that are naturally hornless, or "polled."

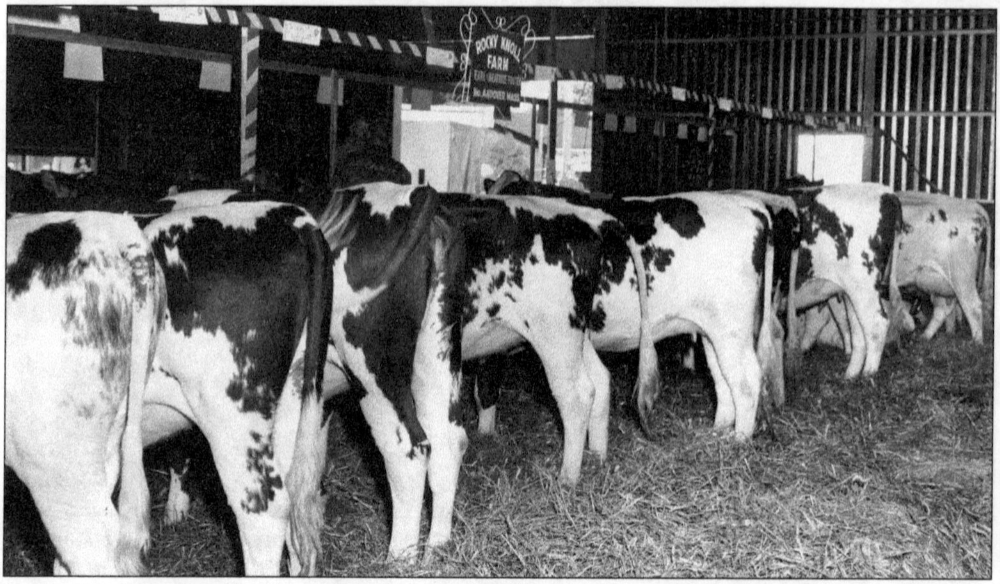

Rocky Knoll Farm in North Andover was where the Fosters housed their prizewinning Ayrshire dairy herd. The Fosters, longtime exhibitors at Topsfield Fair, have their show string in the fair's cattle barn in this photograph from 1950. Earl Foster was a longtime cattle showman at the Topsfield Fair.

The fair patrons were thrilled to see this newly born Holstein calf born in the cattle barn at the start of the fair on Saturday. Yes, mother and baby are obviously doing fine here.

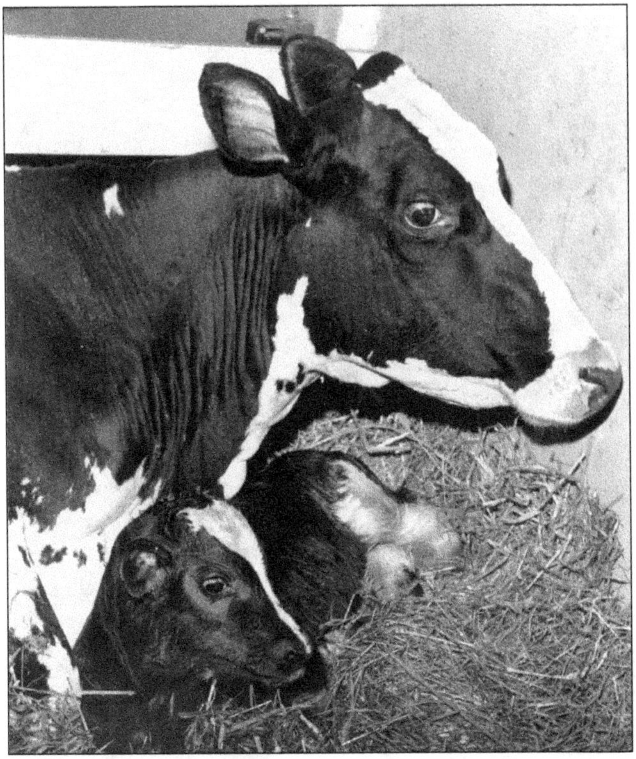

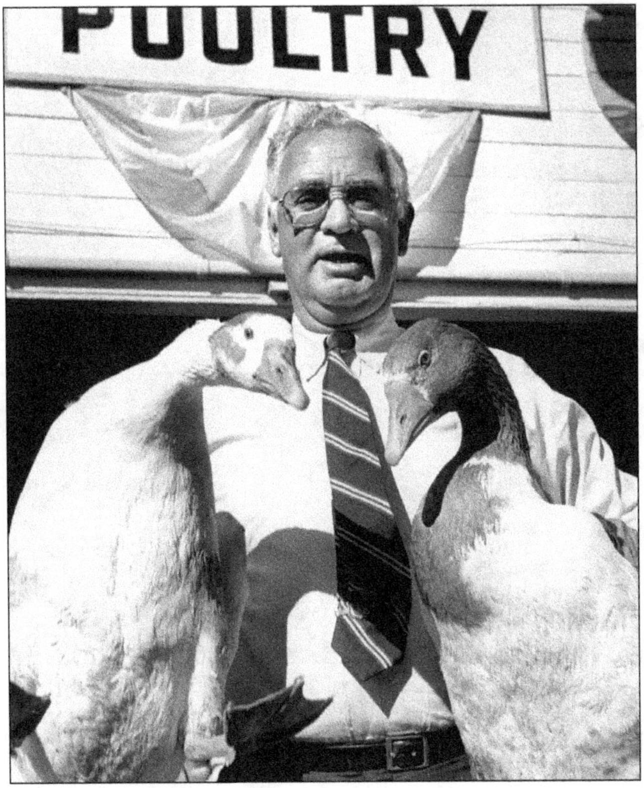

All types of birds are shown at the poultry show at the Topsfield Fair. In 1994, Forrester A. "Tim" Clark of Hamilton displayed his prizewinning geese in the show. Poultry shows have always been a major attraction at New England fairs.

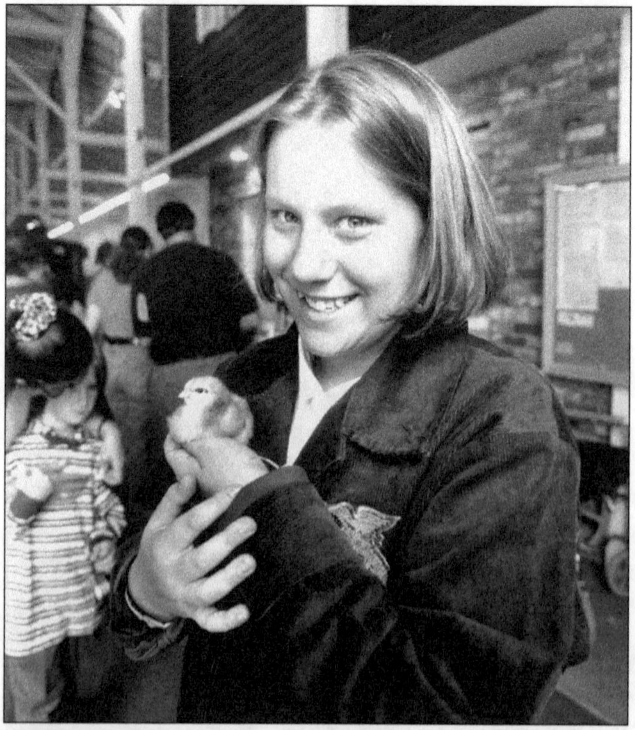

Tiffany Grant of Peabody, an FFA member from the Essex Chapter FFA at the Essex Agricultural and Technical Institute in Hathorne, holds a newly hatched chick from the fair's hatchery in the poultry building, shown here during the 1997 Topsfield Fair.

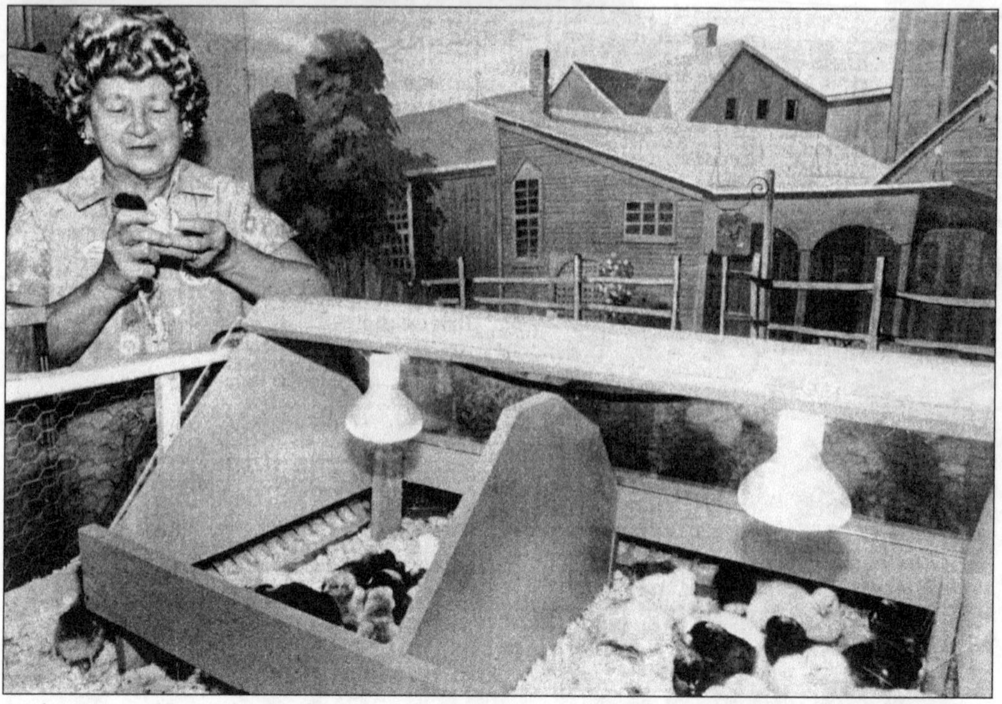

Sophie Jaros of Ipswich affords tender loving care to these baby chicks at the Topsfield Fair in 1986. The hatching exhibit is an ever popular attraction, especially to young fairgoers. Sophie and her husband, John (who still volunteers), were longtime members of the poultry show committee at the fair.

There goes "the Ringer!" At the annual Hen Flying Contest, Tony DiPaolo throws up the hens of each exhibitor to see how far each will fly. The one that flies the farthest is the winner. As the tension mounts during the contest, DiPaolo has been known to slip in a hen pheasant, and when he tosses her into the air, off she flies out of sight and the crowd is amazed. The ruse works every time he does this to astound the crowd. In Beverly, he raises many kinds of exotic birds, including pheasants, grouse, ducks, bantams, and finches.

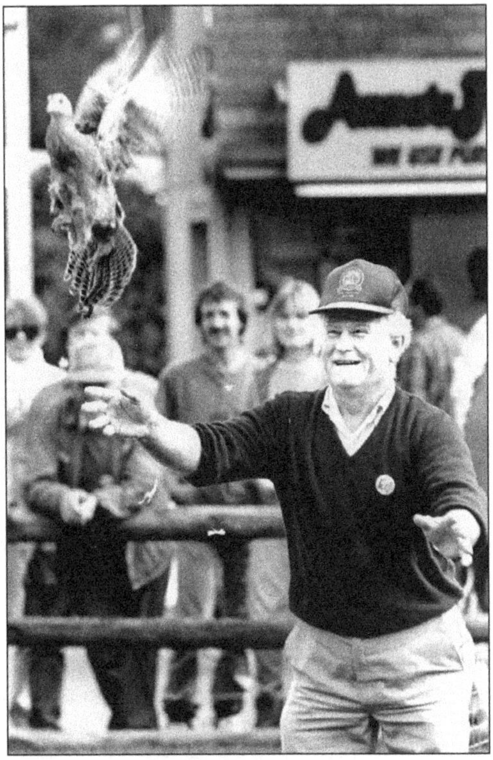

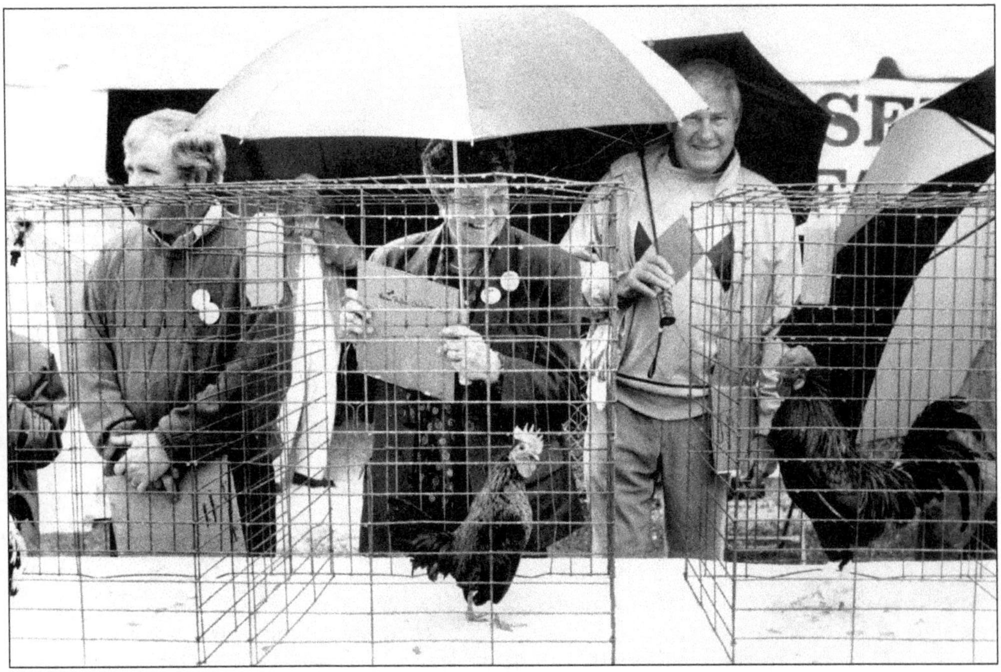

On a very rainy Saturday at a recent Topsfield Fair, John Cole of Topsfield, from the fair's executive board, and Sally O'Maley of Essex, the fair's director of advertising and public relations, stood in the pouring rain to judge the annual cock-crowing contest. The crows in a given period of time are counted to select the winners.

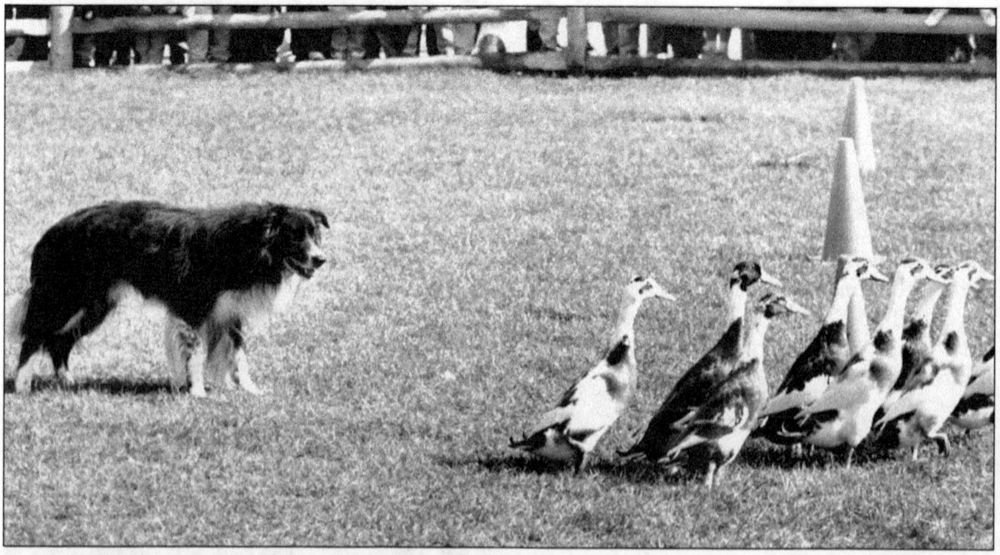

Border collies are natural-born herding dogs, as they instinctively try to round up anything and control its movements, including cattle, sheep, children, and even ducks, as this dog is doing in a 1995 demonstration at the Old Essex County Barnyard. The ducks are Indian Runners, which have this more upright posture and are known as good egg producers as far as laying ducks. With some training to act on voice, whistle, and hand signals, one of the dogs can move a flock of ducks or other animals almost anywhere and into almost anything. The dog here is herding these ducks through an obstacle course and then finally into a small pen.

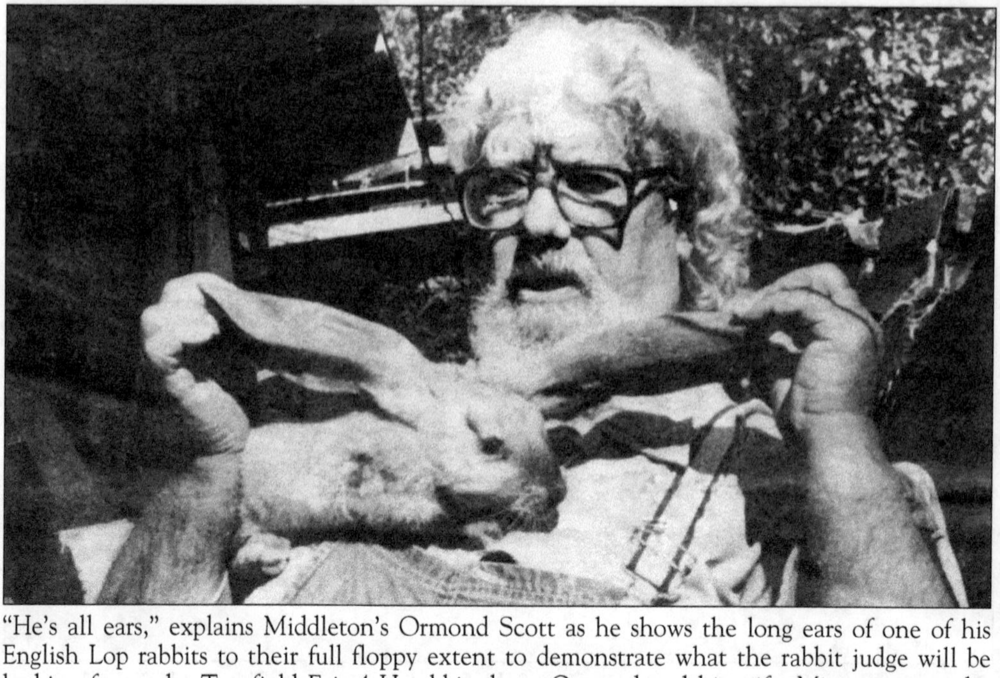

"He's all ears," explains Middleton's Ormond Scott as he shows the long ears of one of his English Lop rabbits to their full floppy extent to demonstrate what the rabbit judge will be looking for at the Topsfield Fair 4-H rabbit show. Ormond and his wife, Minetta, were the leaders of the largest 4-H rabbit club in all of the United States, with well over 125 members. His club is known as the Great Scott 4-H Rabbit Club and still is going strong today with club activities for youngsters interested in rabbits.

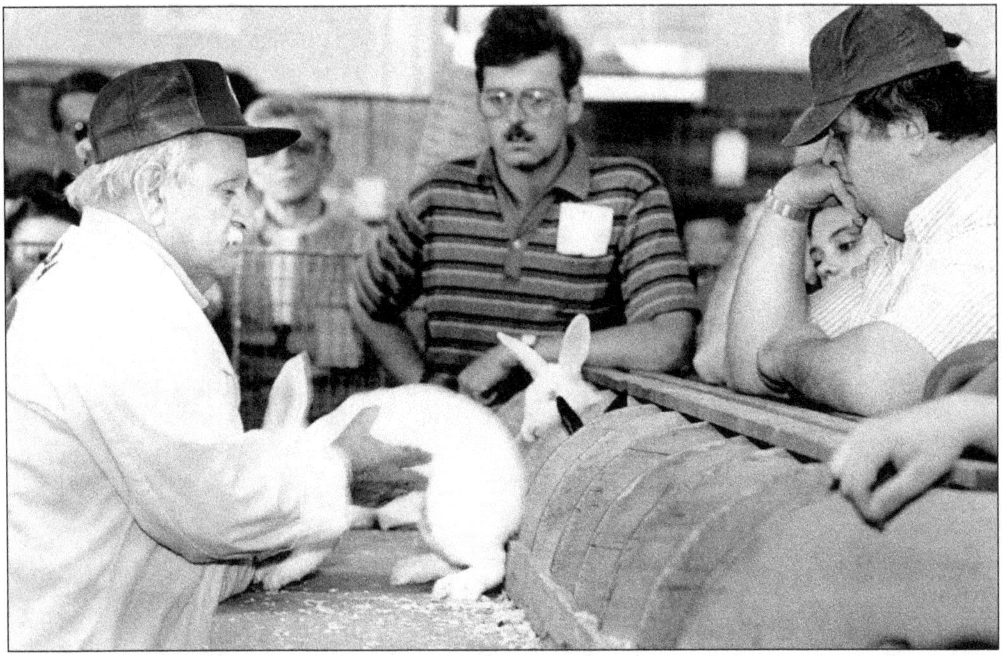

Longtime rabbit judge Manny Cavacco of Kingston makes a difficult decision in the Best of Six Breeds class at the Topsfield Fair. The rabbit building at the fair also accommodates cavies, or guinea pigs.

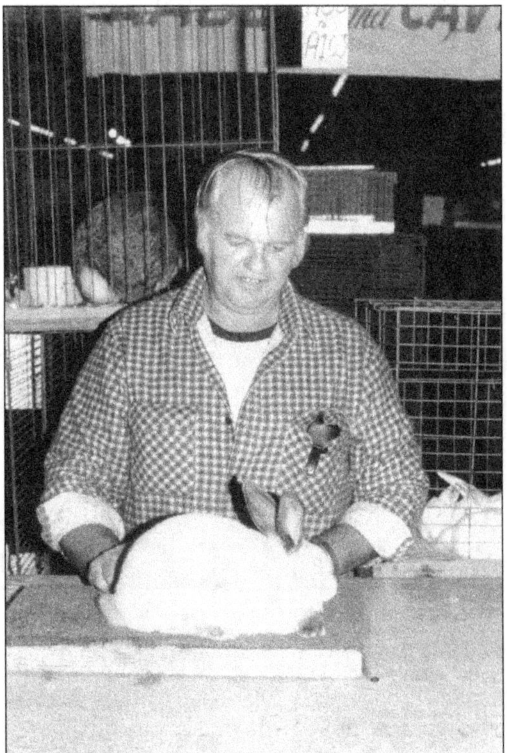

Lawrence Swift checks one of the rabbits in the Topsfield Fair rabbit building. He was the superintendent of the show until 2001.

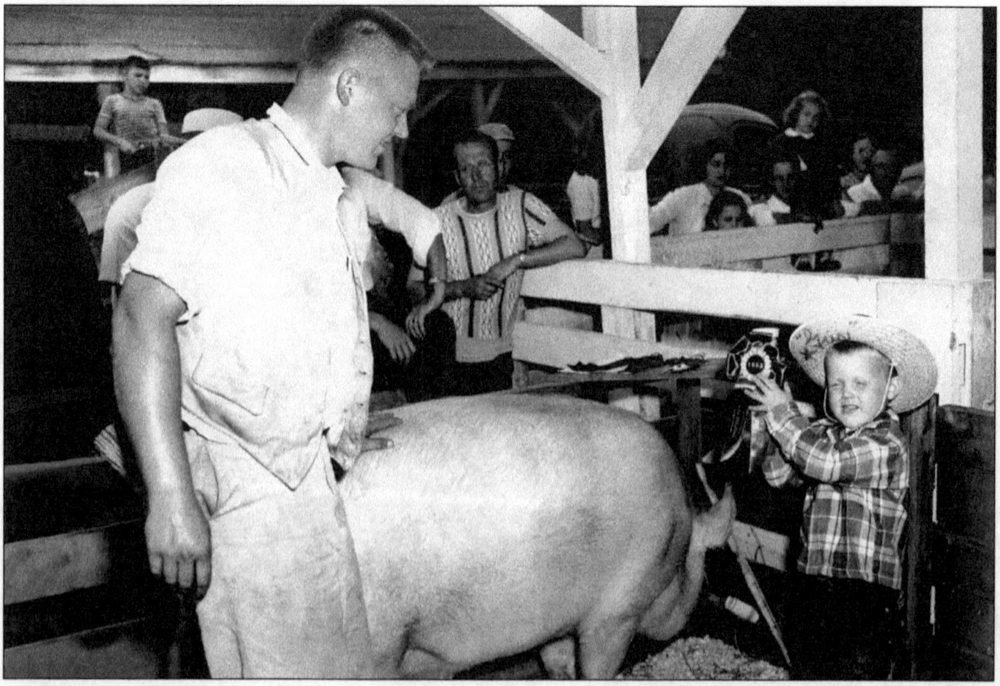

In 1953, Augustus Means, known as Gus, shows pride as his son, "Little Gus," shows off a big blue rosette he has won for his hog exhibited at the fair. Most agricultural fairs also have many sows with litters of piglets to attract the many youngsters to the pig barn. It is great to see these very young exhibitors showing animals and other exhibits at fairs under the guidance of parents or 4-H or FFA club leaders. Both organizations strongly promote agricultural competitions for young people.

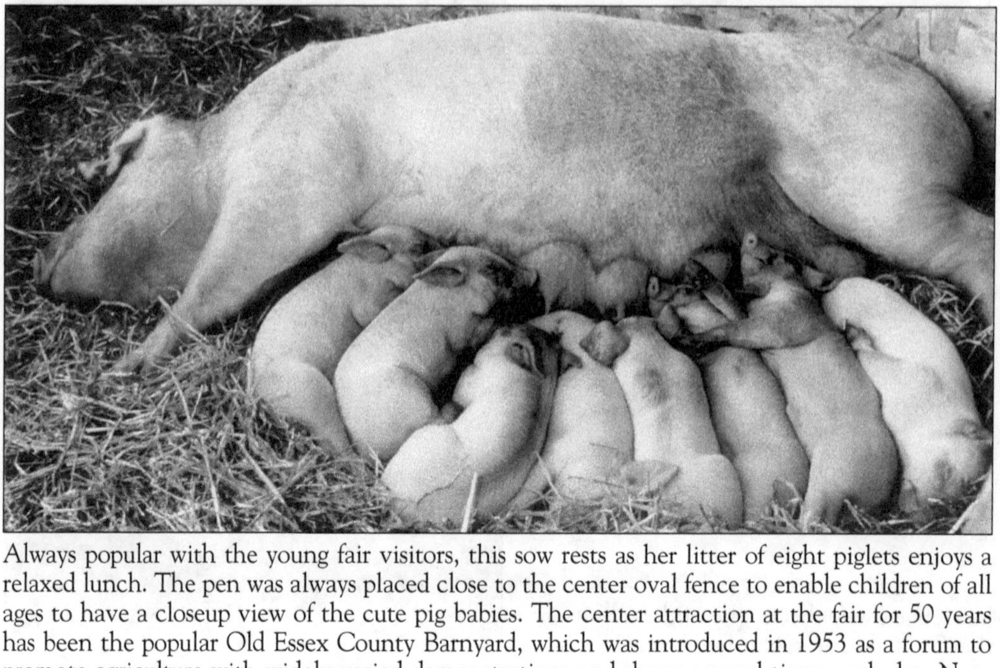

Always popular with the young fair visitors, this sow rests as her litter of eight piglets enjoys a relaxed lunch. The pen was always placed close to the center oval fence to enable children of all ages to have a closeup view of the cute pig babies. The center attraction at the fair for 50 years has been the popular Old Essex County Barnyard, which was introduced in 1953 as a forum to promote agriculture with widely varied demonstrations and shows several times each day. Note that the "Barnyard" was called the "Essex County Farm" in the 1950s.

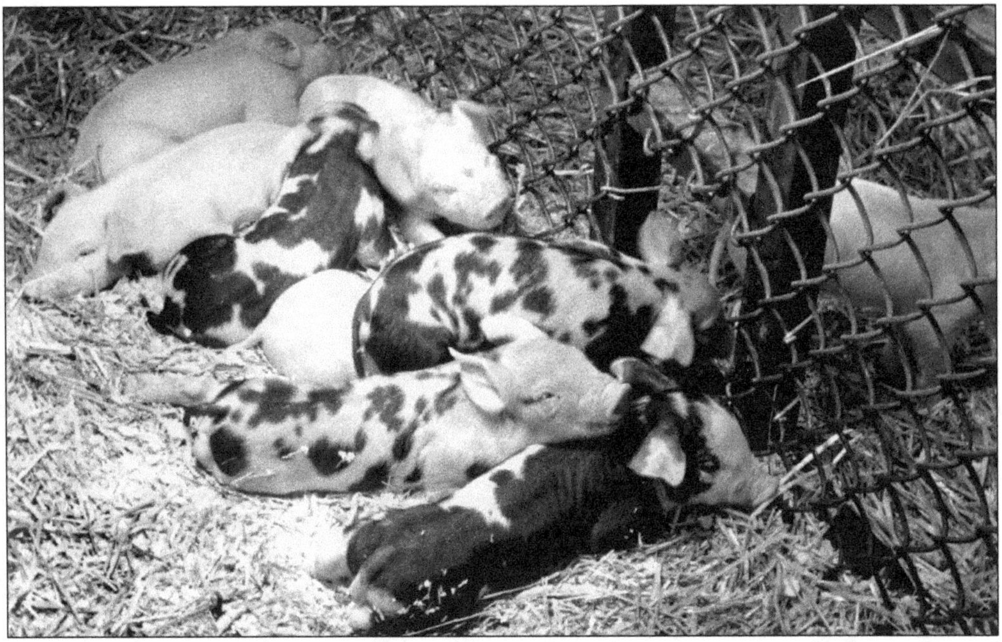

A litter of cute piglets, most of them heavily spotted, slumbers in a pen near the fence of Daniel Hurld's Old Essex County Barnyard in 1995. Note the opening under the chain-link fence to enable the piglets to escape from their mother's weight if she accidentally traps them under her. Many tiny piglets are accidentally smothered by their mothers in small quarters.

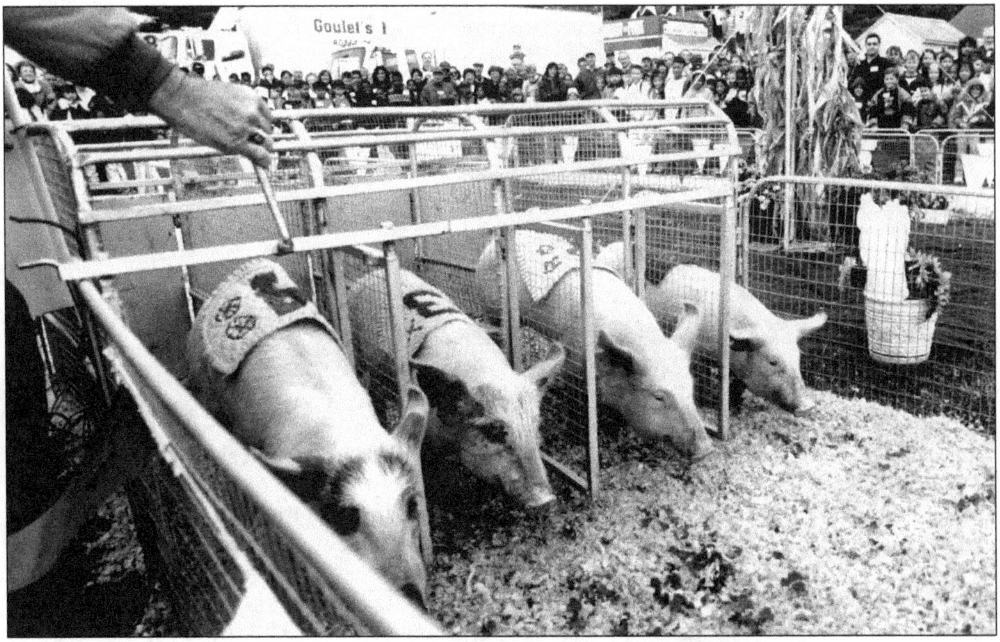

Many kinds of animals have been raced over the years at the Topsfield Fair, but none so cute as the little piglets at Robinson's Racing Pigs racetrack near the 4-H building. These piglets get a well-deserved short run around the tiny track for a reward of Oreo cookies. They perform for the delight of both children and adults alike as they take less than two minutes from start to finish.

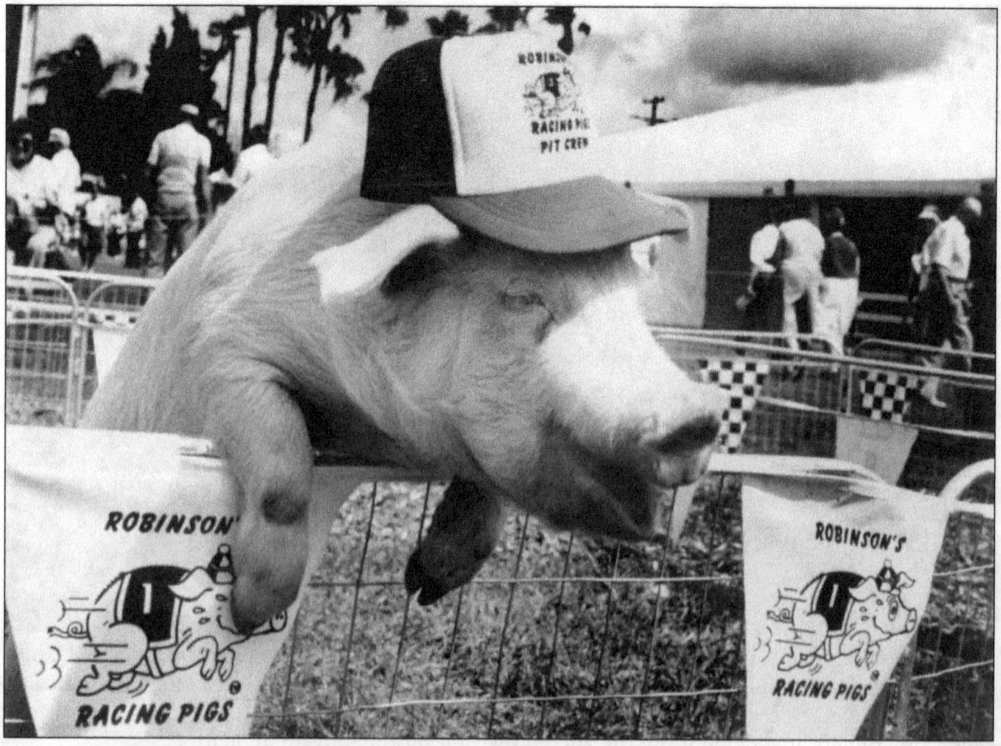

One of the Robinson's Racing Pigs pit crew looks for attention from fans here in front of his trailer, or maybe he has just spotted a unattended Oreo cookie on the ground outside the fence.

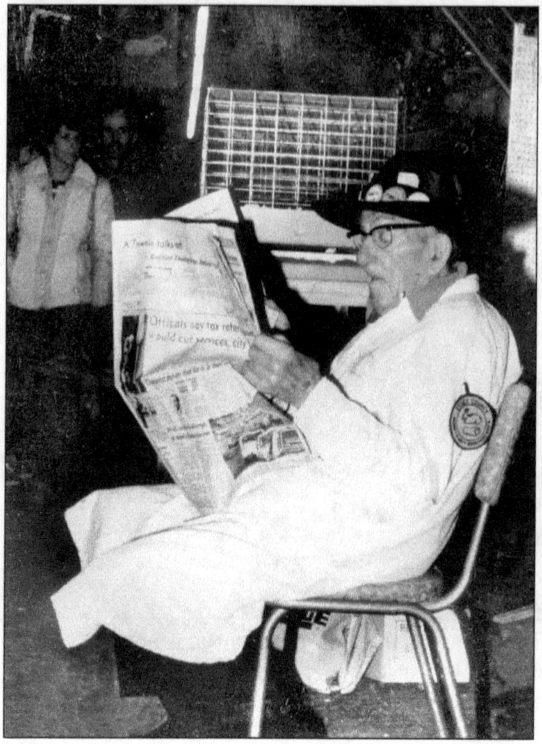

Rudy Fischer of Peabody, longtime superintendent of the rabbit show, relaxes from his duties in the rabbit department. He was associated with the Topsfield Fair for over 60 years, as he started coming to the rabbit shows in the early 1930s. In 1974, the fair presented him with a special citation recognizing his many decades of service to that department. (Photograph courtesy of Roberta Swift.)

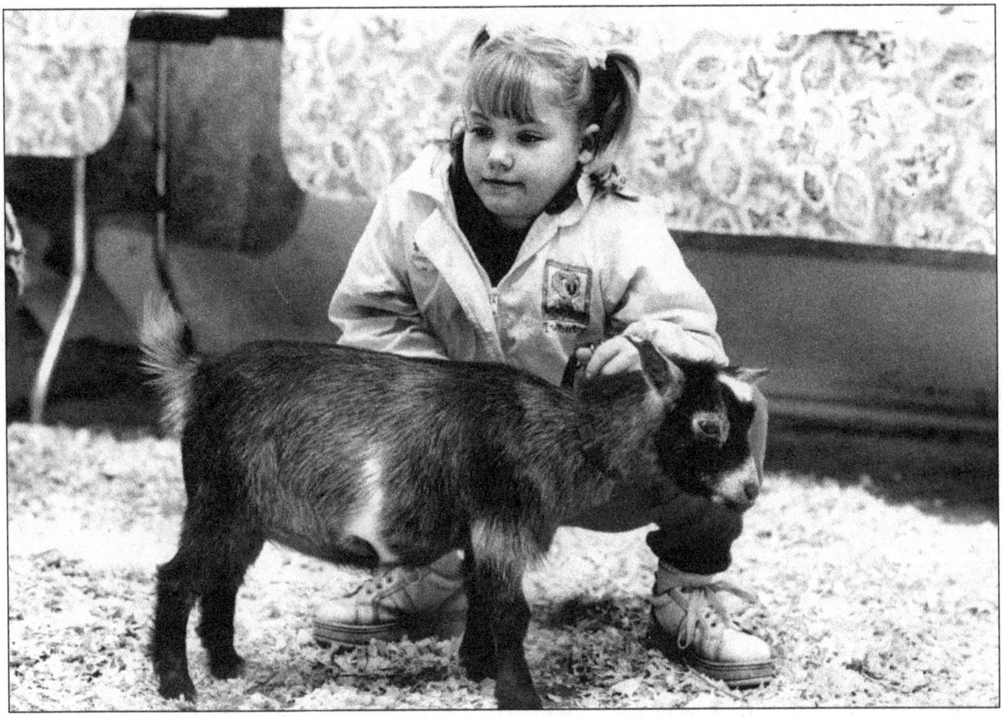

Nine-year-old Hillary Sahm of Phoenix, New Hampshire, is shown with her two-year-old pigmy goat at the Topsfield Fair in 2000. Goats of many breeds have been shown at the fair for many years, especially the dairy goat breeds. Pigmy goats make great pets for children.

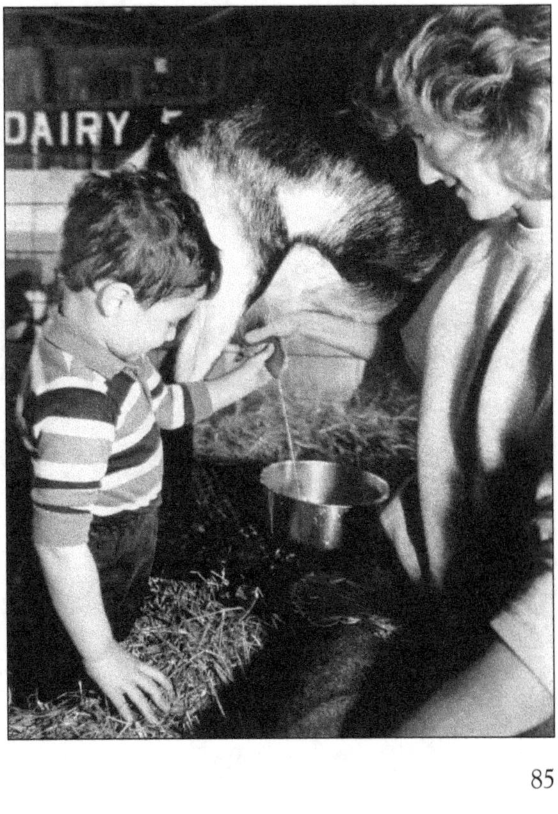

Jennifer Miles of Lynnfield gives a goat-milking lesson to three-year-old Danny Dudley of Salem, New Hampshire, at the fair in 1990.

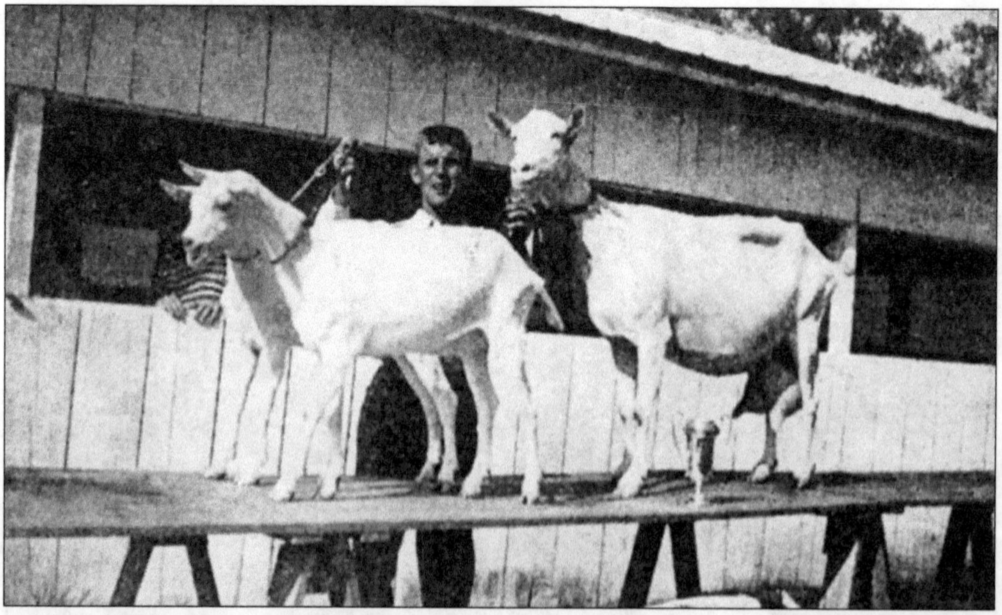

In 1941, the official show of the Massachusetts Council of Goat Breeders Association, sponsored by the Essex County Milk Goat Breeders Association, was held at the Topsfield Fair. In this view outside the goat building, two dairy goats are being readied for the show. The chairman for this show was V. Byron Bennett of Ipswich. Bennett taught mathematics at the Essex Agricultural School for many years.

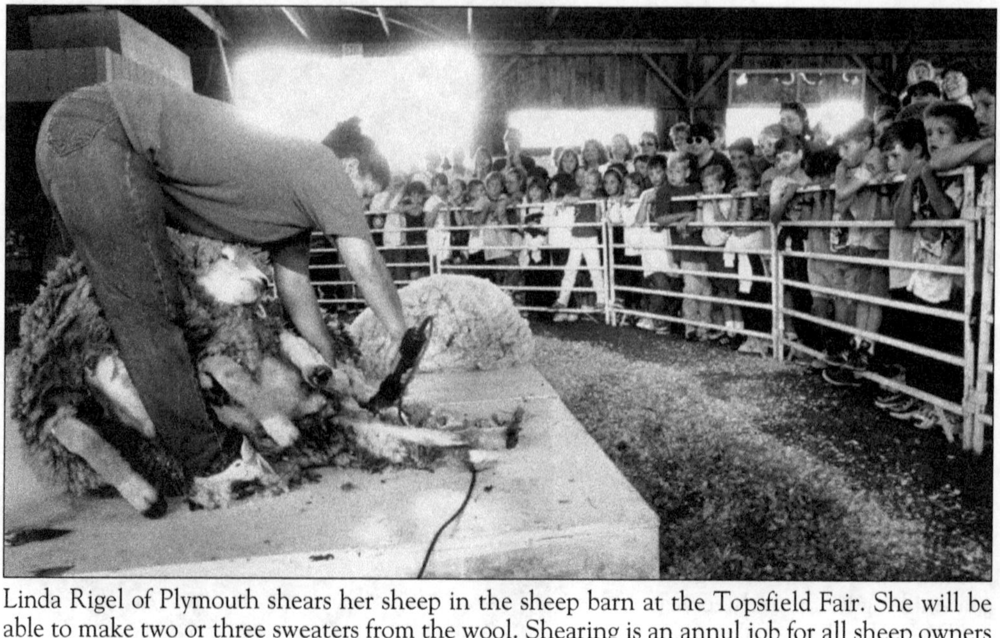

Linda Rigel of Plymouth shears her sheep in the sheep barn at the Topsfield Fair. She will be able to make two or three sweaters from the wool. Shearing is an annul job for all sheep owners to gather the wool fleeces, but the process of shearing can be a backbreaking job.

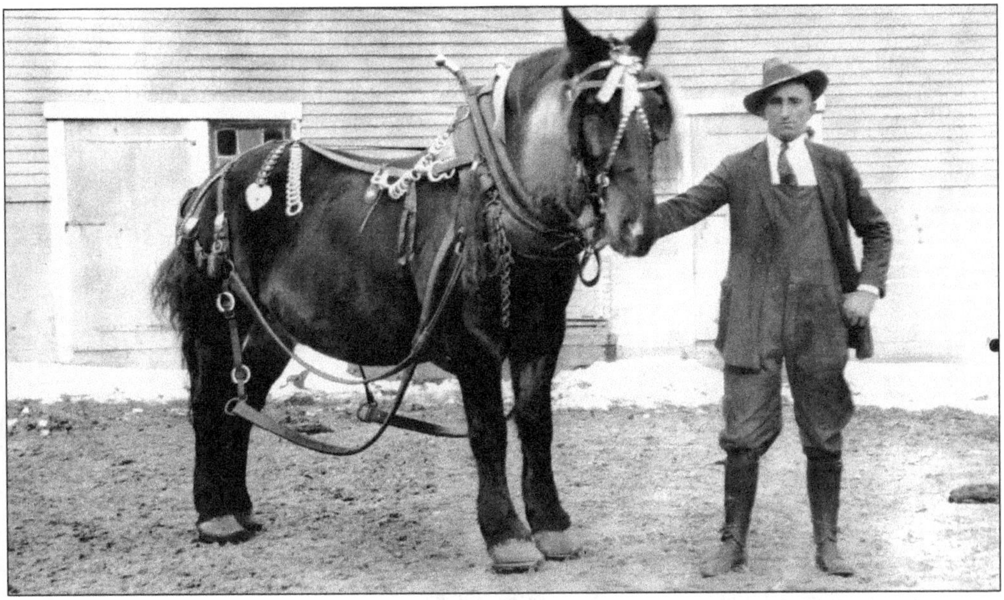

This black Percheron (named Dick) from the Pingree Farm in Topsfield broke the state dynamometer contest at the Topsfield Fair in 1927. In the early pulling contests, the animals pulled against a truck-mounted dynamometer instead of pulling a stone-loaded "boat," or drag, as in more recent years. The record-breaking event was so significant that this photograph was used on the cover of the 1927 *Transactions* report of the Essex Agricultural Society.

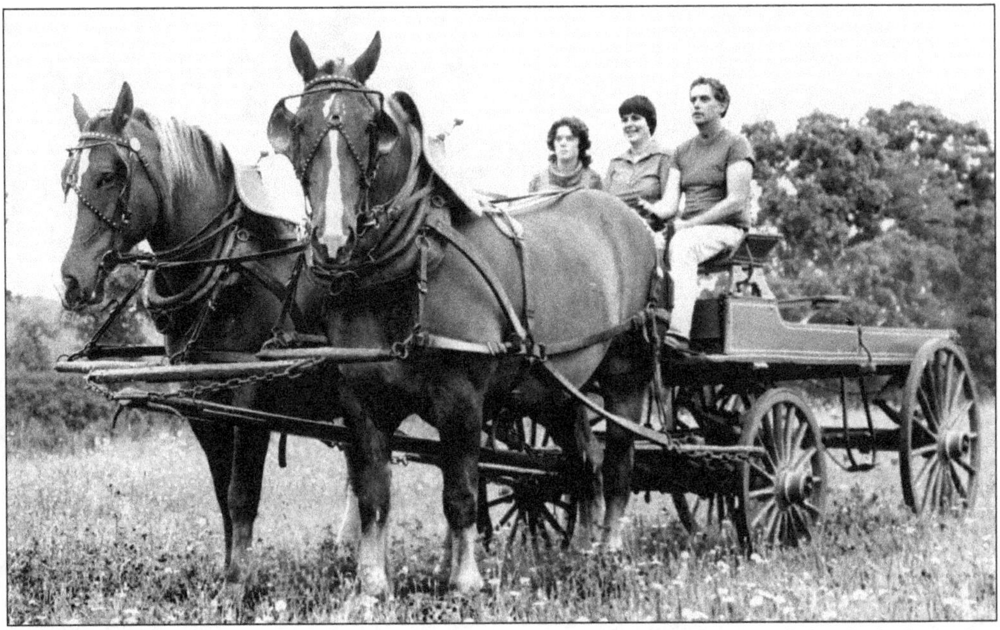

An avid draft horse owner, Alvin Craig (the present general manager of the fair) drives his favorite team of Belgians, readying them for the driving competitions in 1979. He is accompanied by his wife, Caroline, and daughter Lisa, who also drives the team. Some years before Alvin became the general manager of the fair, he was responsible for beginning the great draft horse shows held in the new livestock arena each season. The building of the livestock arena was one of the most important projects that he has helped make happen.

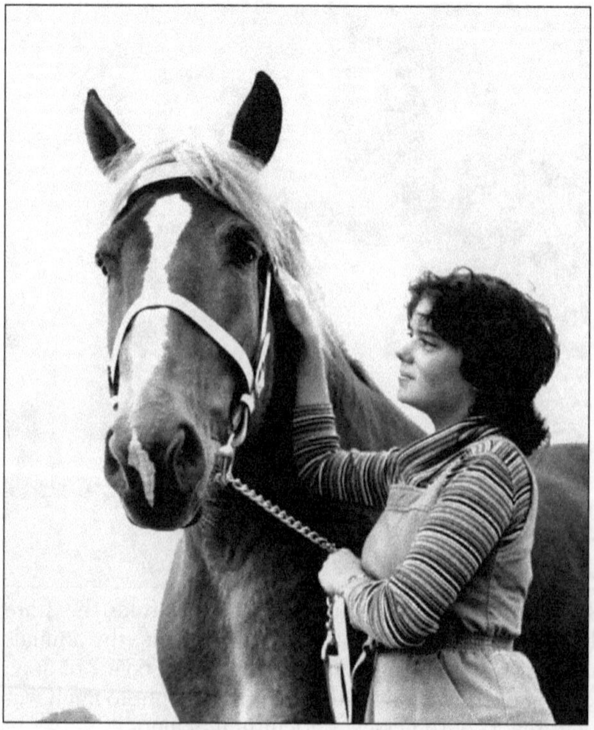

In this 1979 view, Lisa Craig shows off her favorite Belgian draft horse, Bill. The horse is one of a heavy pair of Belgian horses owned by her father, Alvin Craig, general manager of the Topsfield Fair since 1987 and founder of the first modern-day draft horse shows at the fair.

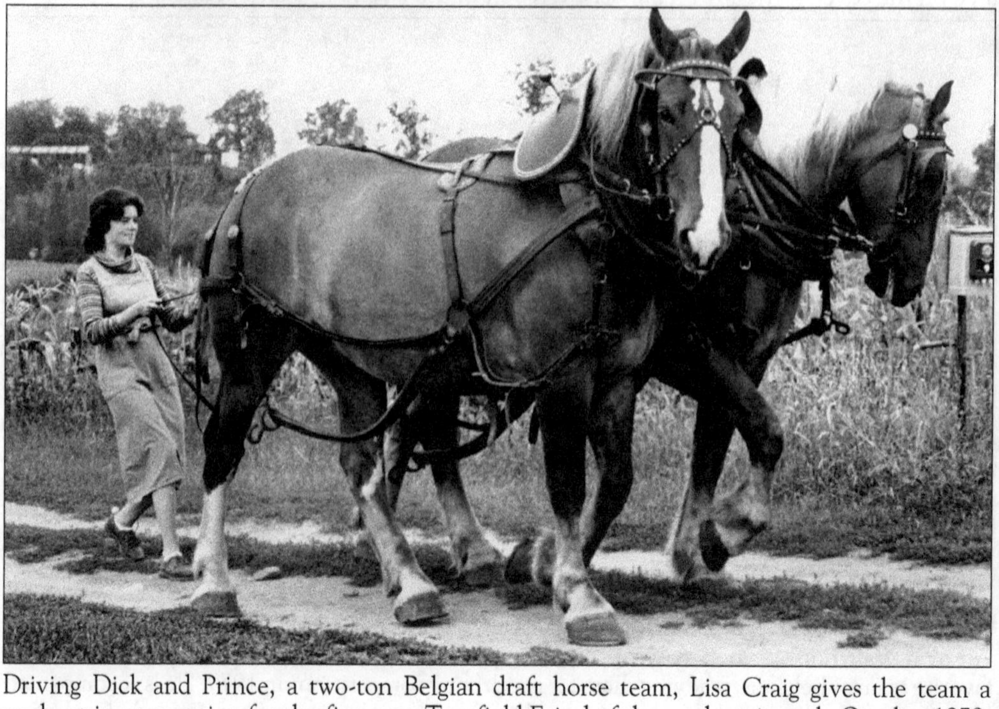

Driving Dick and Prince, a two-ton Belgian draft horse team, Lisa Craig gives the team a workout in preparation for the first-ever Topsfield Fair draft horse show in early October 1979. The show was held on October 13 during that year, with many fine teams of heavy horses competing. Lisa, an accomplished horsewoman, is the daughter of Caroline and Alvin Craig of Essex.

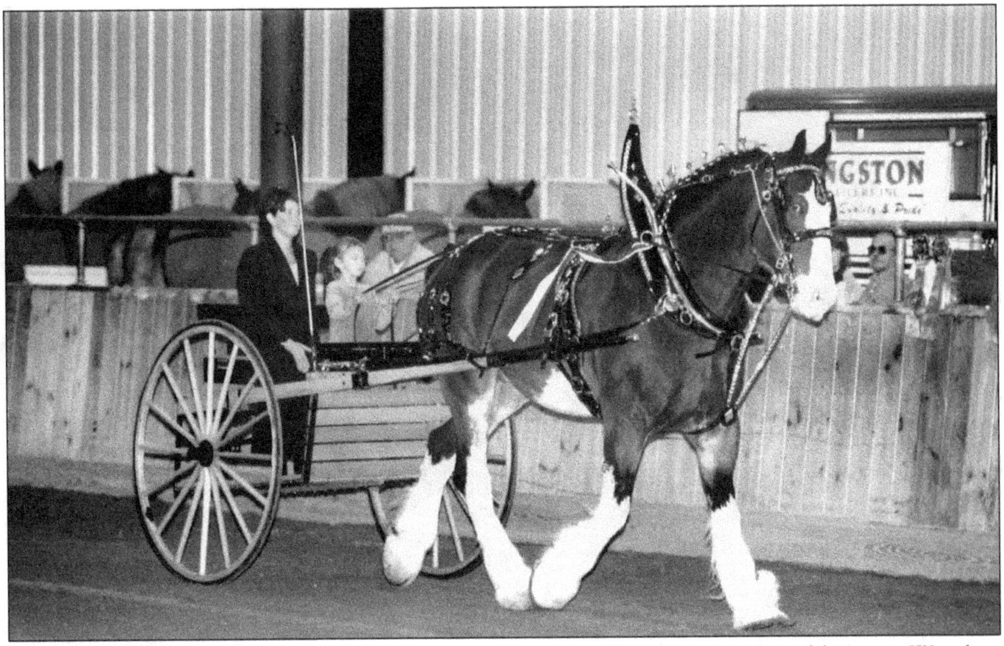

The year is 2002 at the annual draft horse show, and here is five-year-old Anna Winslow from Maine driving a magnificent Clydesdale in the cart class. The horse is owned by Michael Liberty.

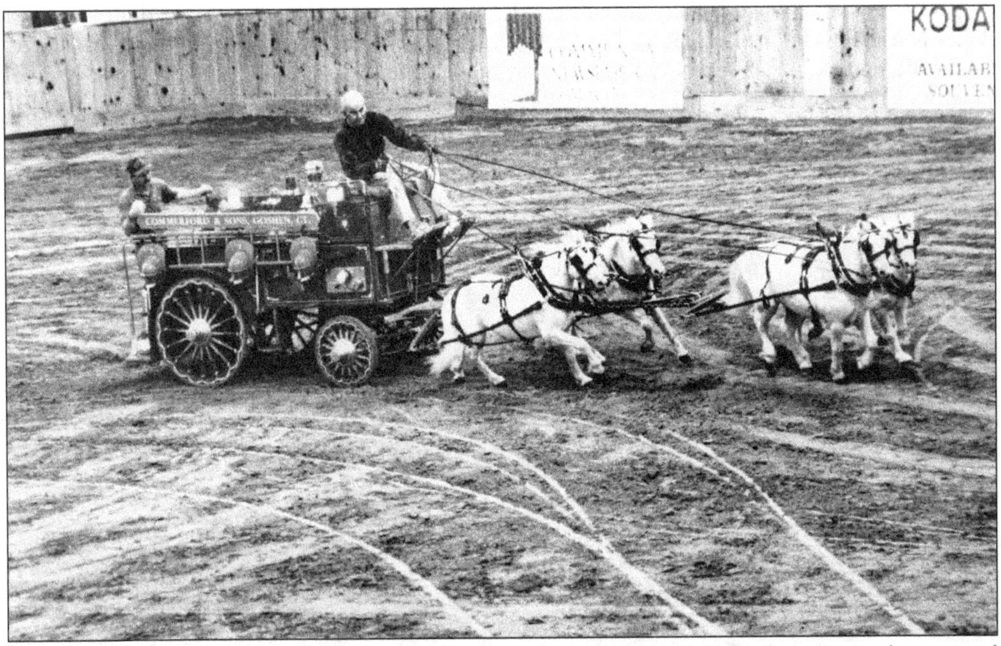

Bob Commerford from Goshen, Connecticut, drives his four-in-hand of miniature horses and to-scale fire engine-wagon through various maneuvers and obstacles at the Topsfield Fair in 1995. The Commerfords also operate the children's petting zoo each year near the flower building to the left of the main gate and behind the sportsmen's building, formerly the 30- by 60-foot dance pavilion (built in 1921).

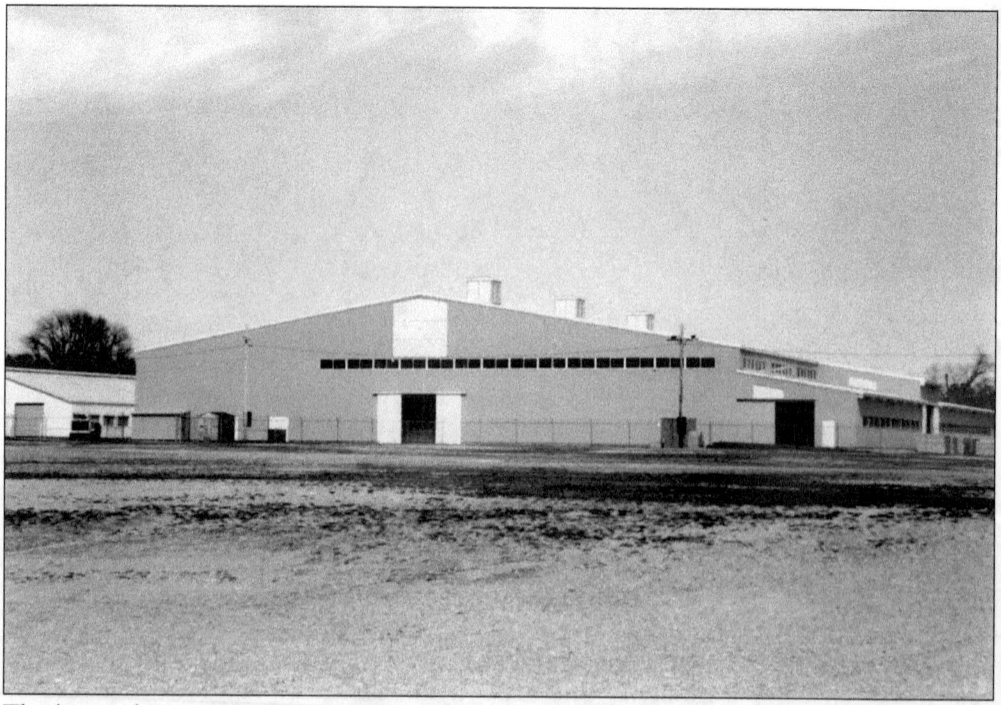

The livestock arena was constructed in 1990 as a new facility and was badly needed to hold all kinds of livestock shows indoors. Now there are major horse and cattle shows held in this facility. Also many horse-driving and riding exhibitions, as well as the Royal Canadian Mounted Police Musical Ride, are held here. The pumpkin contest fills most of the arena on the opening day of the fair. At other times during the year, many groups hold major athletic events in this cavernous building.

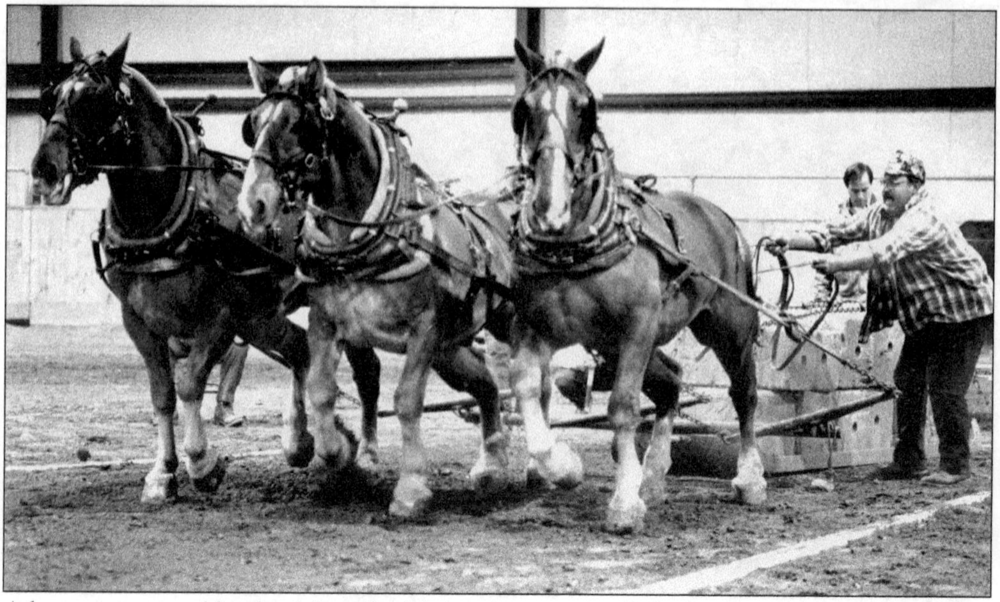

A less conventional hitch of three draft horses strains to pull the load on the boat in the horse pulling event in the new livestock arena at the fair.

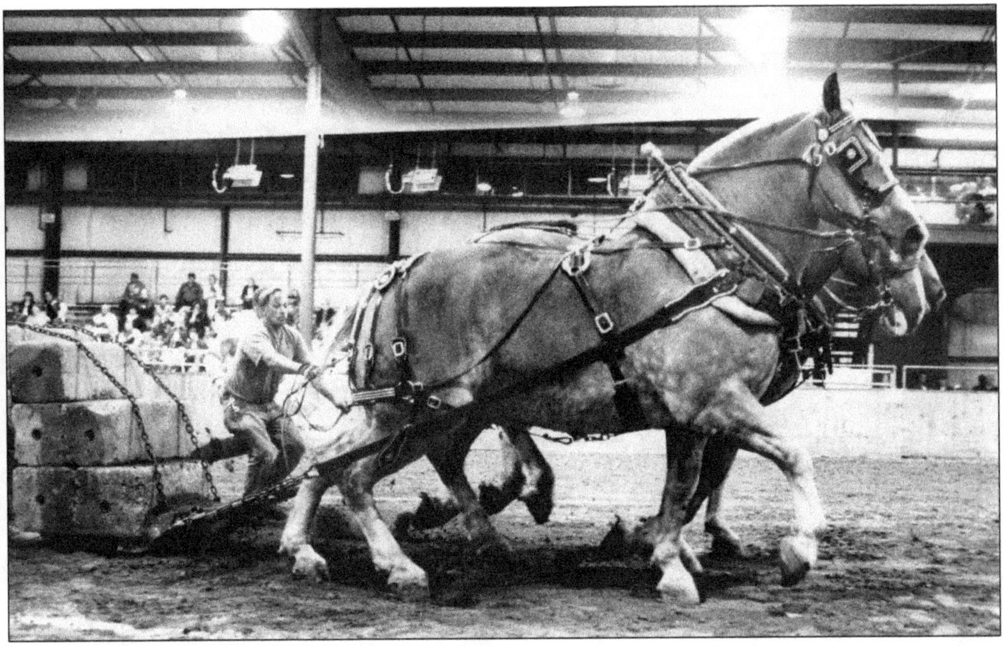

During the 1995 competition, there is a lot of weight on the boat as James Grover from Connecticut urges his heavy team of fine Belgian draft horses forward to achieve the necessary footage to win the match in the new livestock arena.

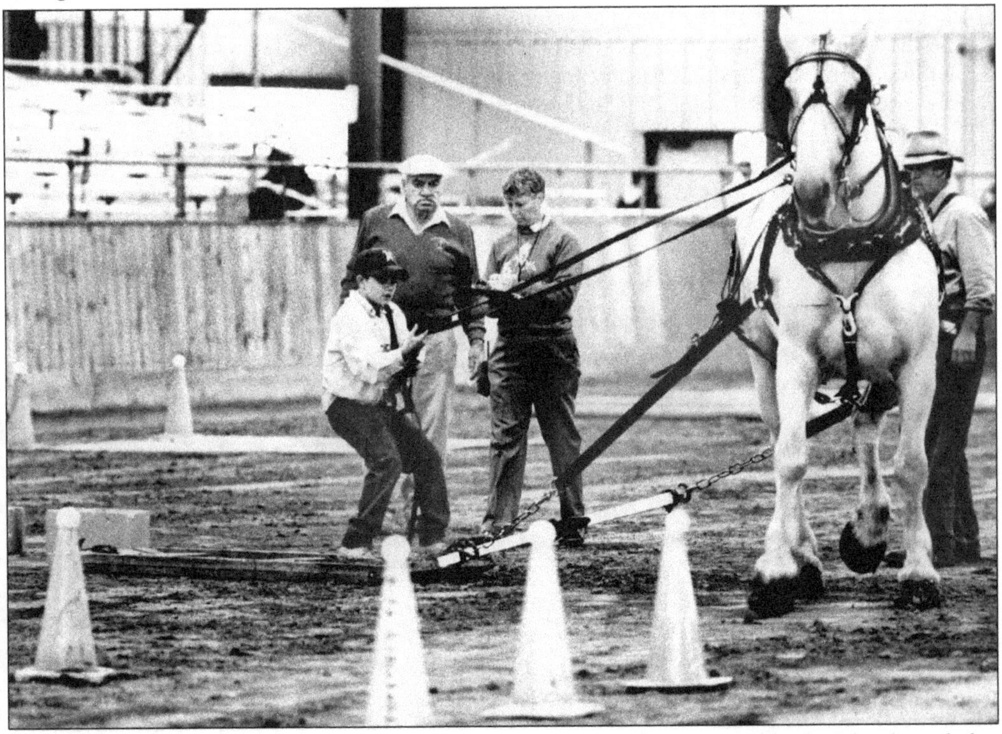

In the new livestock arena in 1995, this young man from Tiverton, Rhode Island, with his father supervising on the right, rides the stone boat as he deftly drives his large draft horse through the obstacle course at the annual draft horse show on the last day of the fair.

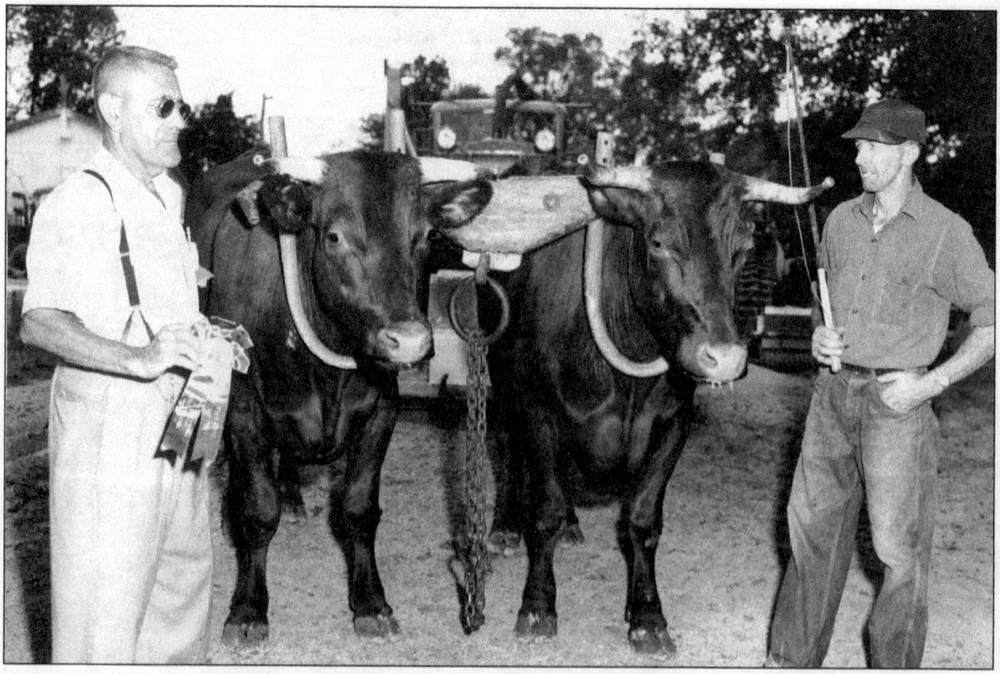

A pair of fine Devon oxen, named Duke and Jerry, have just successfully won their weight class in the oxen pulling contest by pulling 5,020 pounds. Maj. Henry C. Thurlow, chairman of the show, is presenting the first-place rosette to their owner and driver, Frank Pinkham of Rochester, New Hampshire, in 1952. Major Thurlow, a World War I veteran, was chairman of the pulling events at Topsfield Fair for many many years. He was also a history teacher at the Essex Agricultural School in nearby Hathorne.

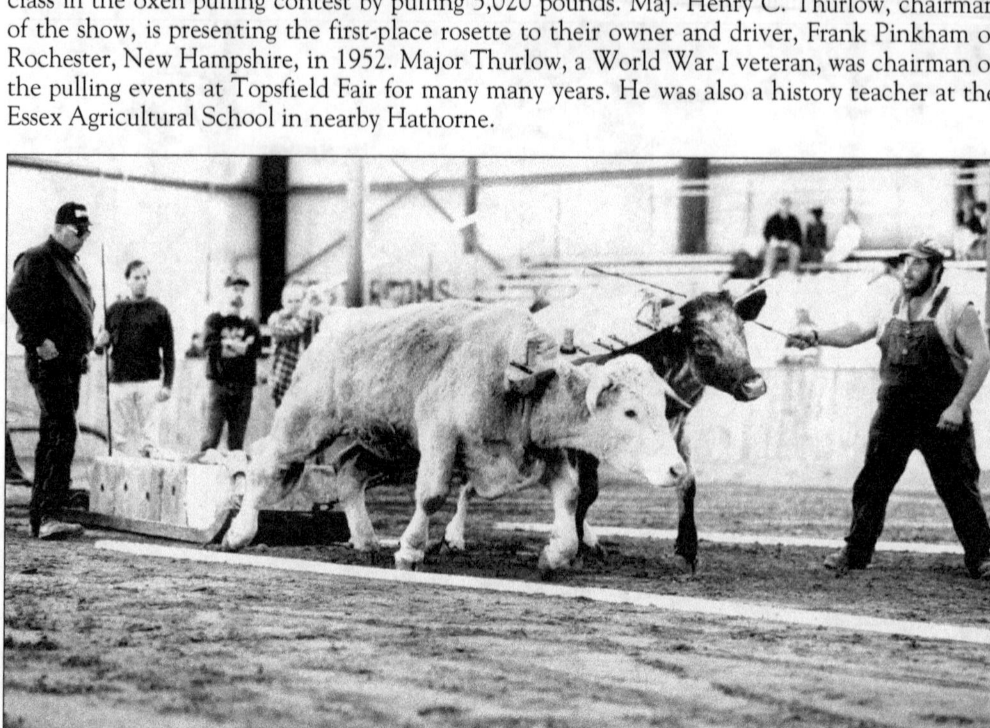

Bruce Lethbridge and Louie Clark are shown as their team of oxen pulls concrete blocks at the Topsfield Fair in 1997. The blocks are loaded onto a boat, and the team has three chances to pull it the required distance in each go-around or they are eliminated from the competition.

Arnold Chappell of Salem, Connecticut, is shown leaving the livestock arena at the Topsfield Fair after competing in the oxen pulling competition recently.

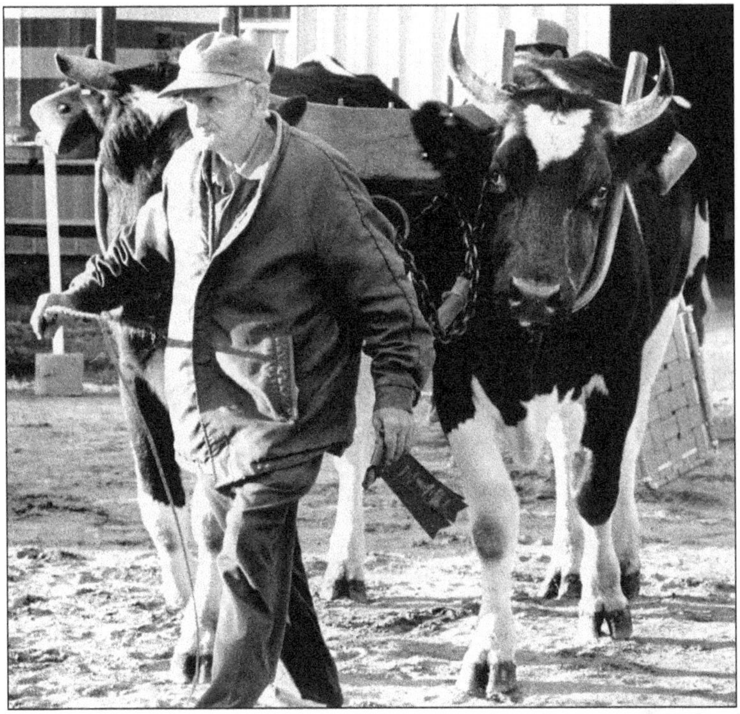

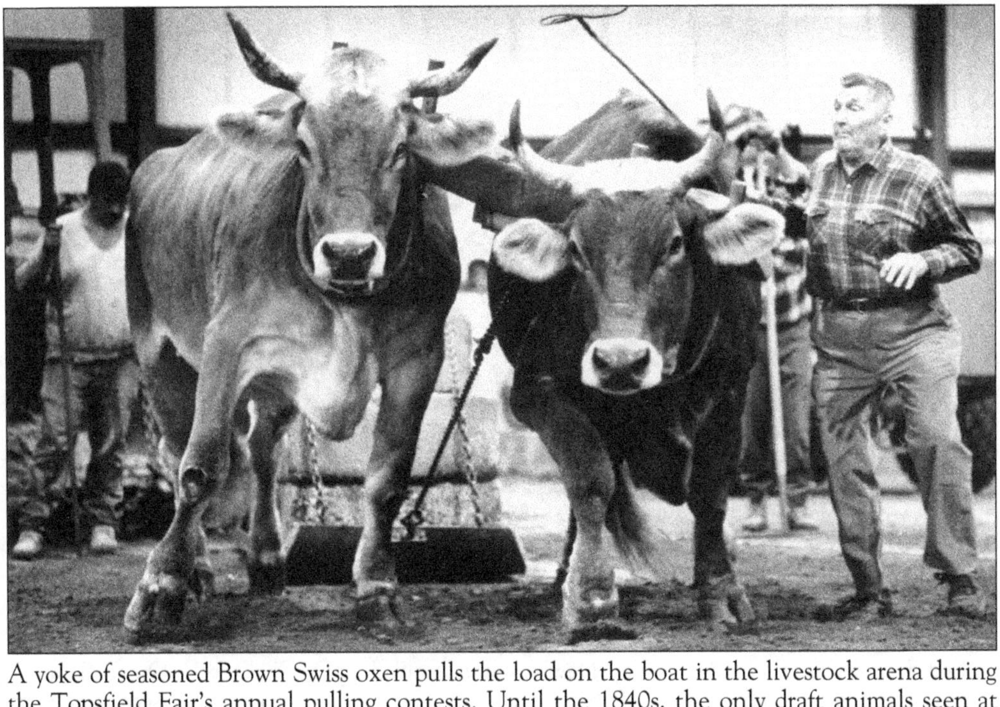

A yoke of seasoned Brown Swiss oxen pulls the load on the boat in the livestock arena during the Topsfield Fair's annual pulling contests. Until the 1840s, the only draft animals seen at the cattle shows and fairs were oxen. Early New England farmers depended on oxen as their principal motive power until the horse gained popularity in the mid-1800s. Horses were more expensive and required more feed and care than oxen did, and they were not a by-product of the farms' milk-producing herd, as the oxen were.

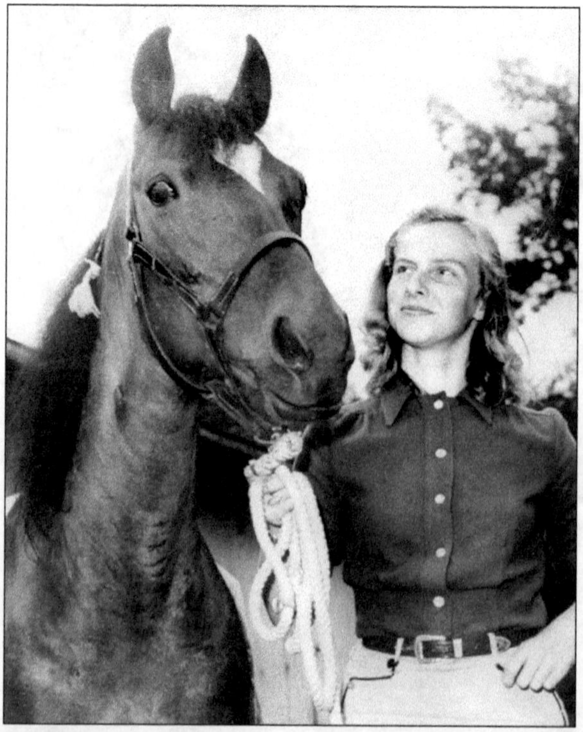

The gas shortage did not bother 16-year-old Barbara Shepard of Stoneham when she decided to go to the Topsfield Fair. Her trusty horse furnished the transportation for the 20-mile journey. Fair officials announced that Shepard was the first person to come to the fair on horseback in 25 years. She said she thought she might have to miss the fair for the first time in seven years because of the gas rationing due to the war, but her horse came to her rescue. She later married Daniel Hurld, and shortly after, the couple moved to West Newbury, where they still reside with their horses. (Photograph courtesy of Barbara and Daniel Hurld.)

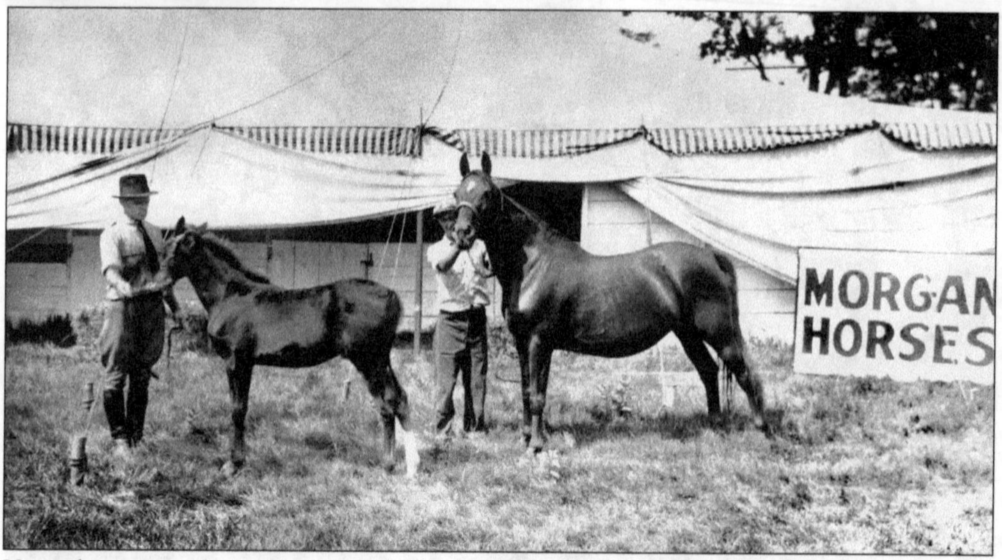

Horse shows were always a very popular feature at the fair. Morgan horses originated in the 18th century in Springfield, Massachusetts, with a little stallion named Justin, named after his owner, Justin Morgan. They continue to be one of the predominant breeds of light horses shown in the United States. Orcland Farms of West Newbury was founded by Dr. Wallace D. Orcutt. Famous Morgan horses were bred, raised, and shown there for many decades. Dr. Orcutt, seen here with an assistant outside the horse tents at the fair, is showing off his prizewinning chestnut mare and foal entered in the breeding mare and foal class at the Topsfield Fair in 1926. The dam is Meh Lady Woodbury, and her foal is Vigiline. This Morgan breeding and showing tradition has been continued on by Wallace's sons, grandchildren, and other family members to the present day.

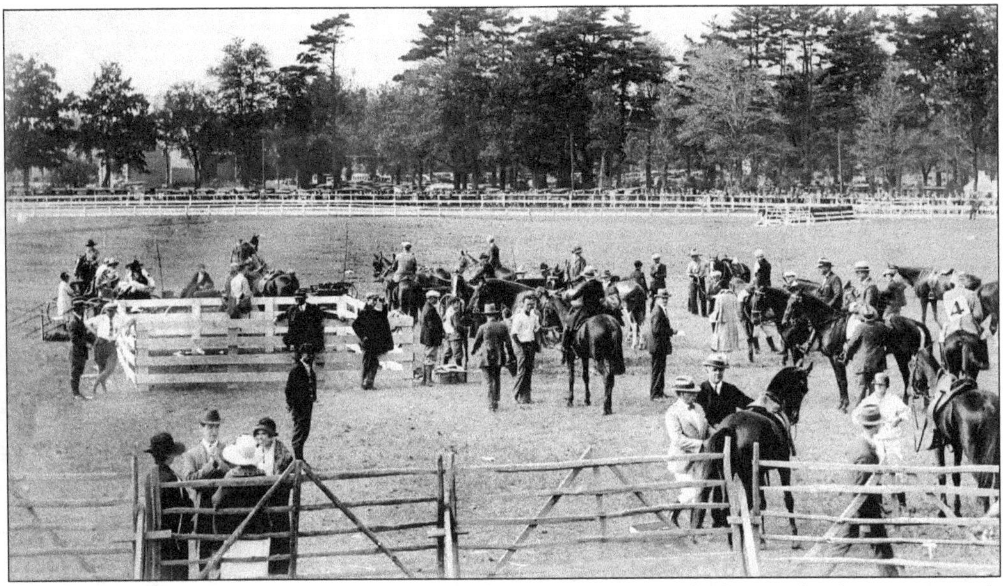

Major horse show events have been held here on the fairgrounds during the fair season each year. Some of the best documented are like this one from 1926. There is one lone male rider in Western attire here in the back of the group of riders. We can see show buggies used for ride and drive classes. According to show programs in the Essex Agricultural Society archives, the West Newbury Riding and Driving Club often sponsored these events at the fair during the 1920s and 1930s.

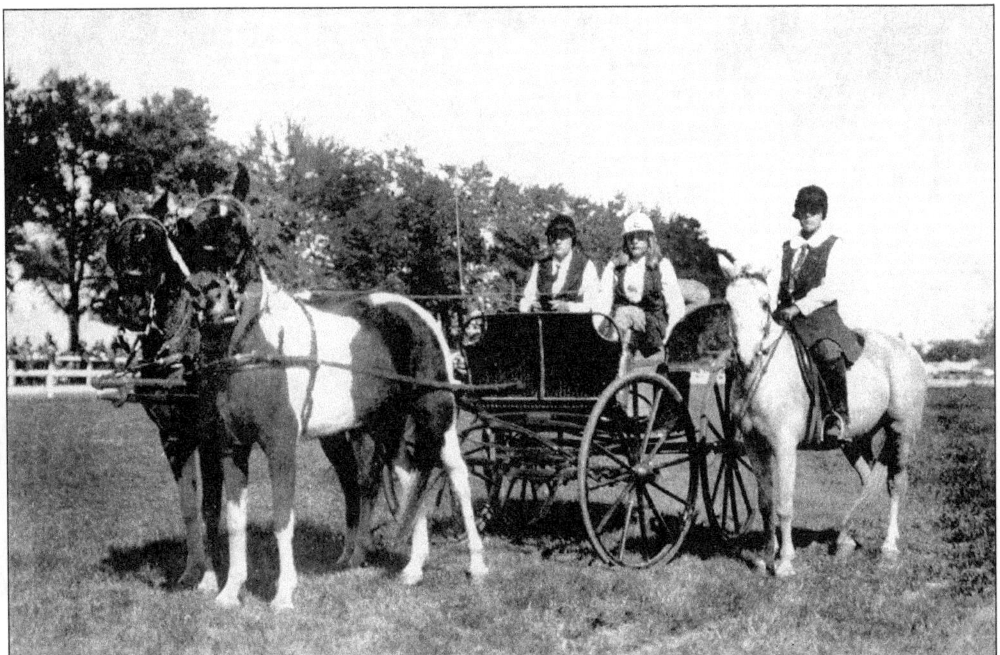

The West Newbury Riding and Driving Club sponsored horse shows and gymkahnas in the 1920s and 1930s. Here we see two ladies driving a matched pair of pinto ponies to a ladies' basket phaeton on the fair's show grounds. They are presumably going to compete in one of the carriage-driving events.

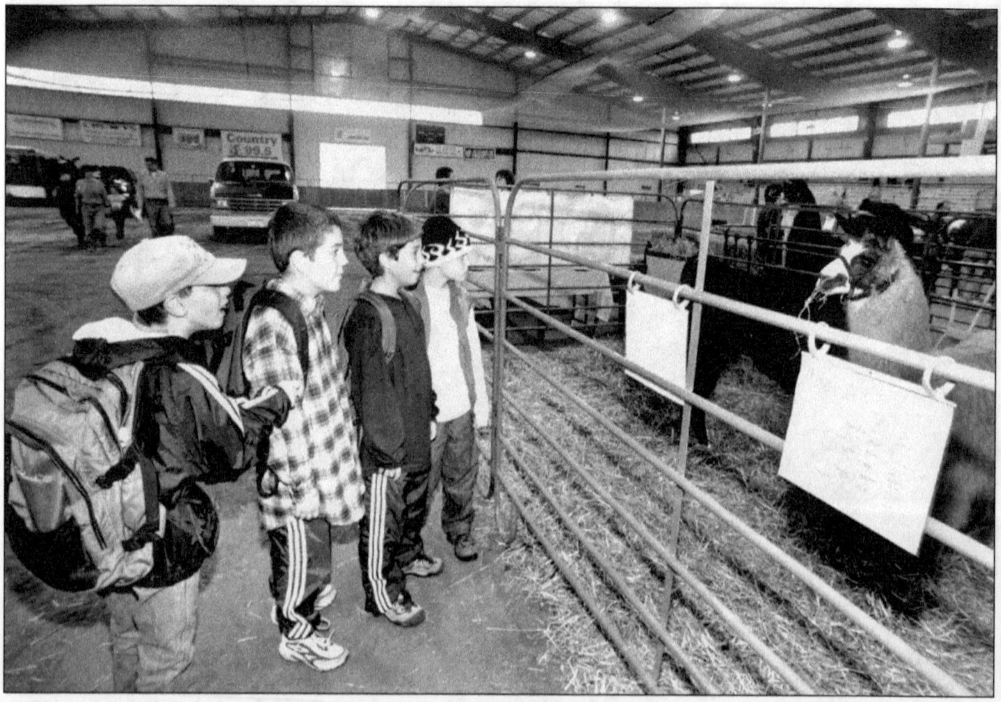

Enjoying the llamas in the livestock arena at the 2000 fair are, from left to right, West Parish School fourth-graders Brian Lloyd, Nicholas Barusso, Joe Borge, and Nicholas Sacharuk.

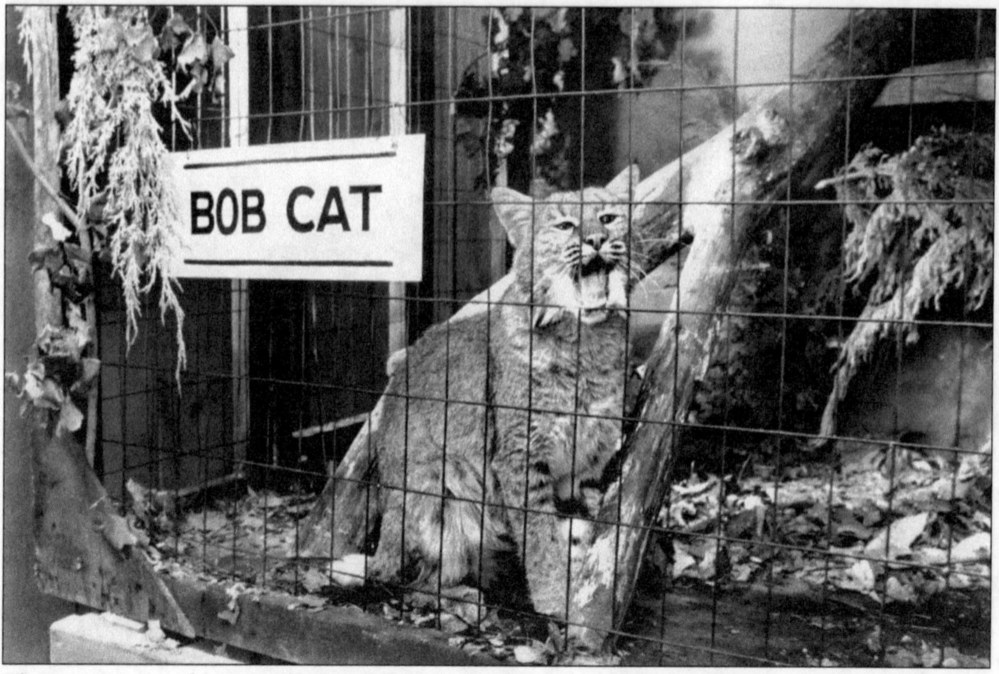

The sportsmen's show usually has some Massachusetts wildlife on display, such as this native bobcat. The bobcat lives mostly on wild rabbits, mice, birds, and whatever else it can find to prey on. They are elusive and are rarely seen during the day, although their numbers in New England have increased in recent years.

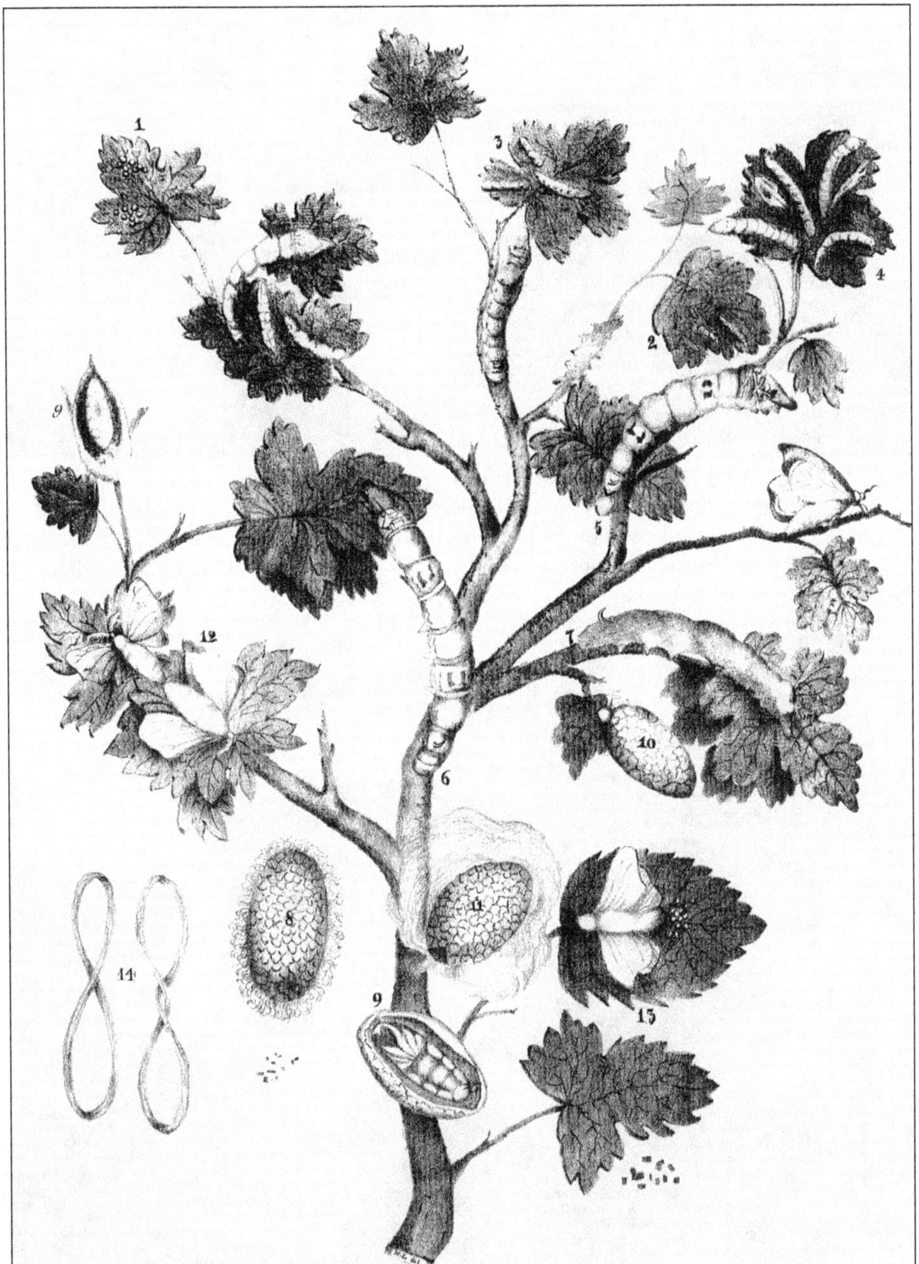

In the first half of the 19th century, there was an increasing interest in raising silkworms in New England. Women, children, and the infirm were encouraged to participate in the activity. The 1832 volume of *Transactions* gives a description of the values of silk production and what was necessary to carry on this endeavor, including this plate, showing the life cycle of the silk moth. There was also a chart showing how many mulberry leaves were necessary to support the worms at various ages in their growth. Enoch Boynton's farm in Byfield had some 42,000 mulberry trees. The industry ultimately failed in Essex County due to severe winters that killed the mulberry orchards, the overpicking of the trees leaves, and the competition by the cotton mills as the textile industry was developed within the Northeast and the industrial revolution progressed.

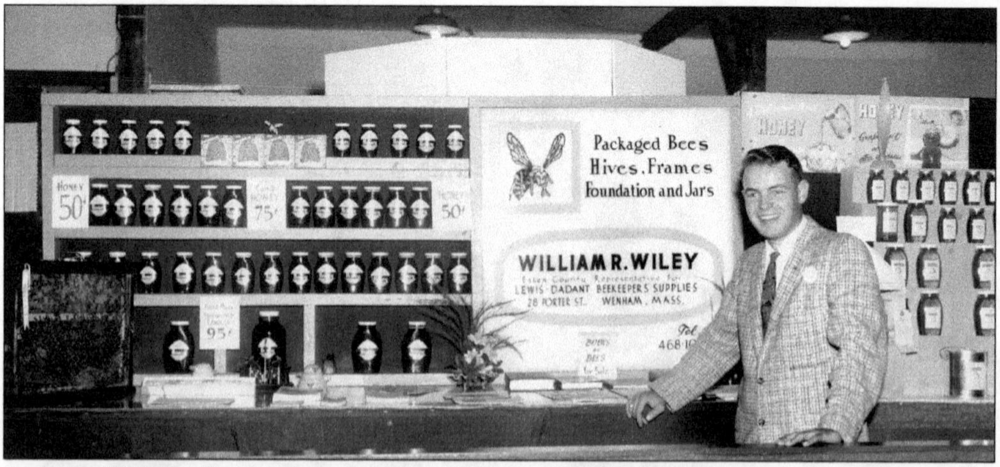

In 1962, William R. Wiley greets fair patrons at his booth in the Grange building at the apiculture department of the Essex County Beekeepers Association. The beekeepers sold honey and beeswax candles during the fair. Wiley was the county representative for the beekeeping supply house of Louis Dadant. Dadant and his company have supplied beekeepers with supplies and bees for more than 75 years.

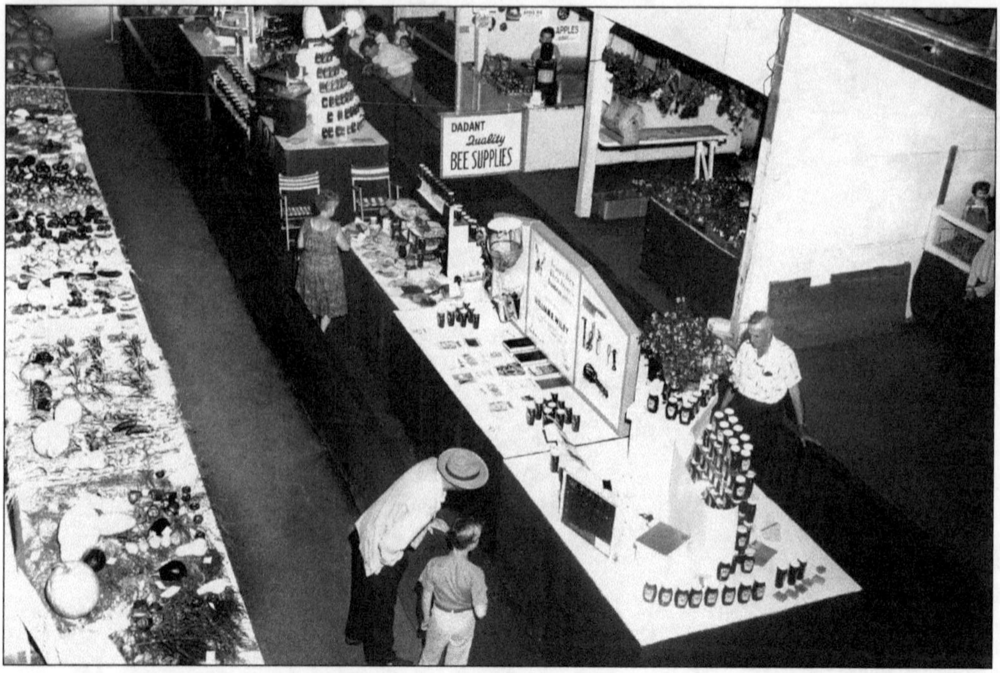

This 1960 view from loft of the Grange building looks down upon the center display area where the various members of the Essex County Beekeepers Association exhibited anything having to do with bees and sold honey and other bee-related products. Martha and Arthur Leach are tending honey sales on this day. On the left are many of the "junior vegetable" exhibits from around Essex County. Halfway down the right aisle is an exhibit of the Essex Chapter FFA from the agricultural school in Hathorne. To the right of the chapter's sign, a student is selling apples and cold cider to fairgoers. All of the apples were grown, picked, and graded by the students at the school. This is the very stand where, in 1970, hot apple cider was first introduced to New England fairs at the suggestion of the Royal Canadian Mounted Police on one cold September night.

Seven

ENTERTAINMENT

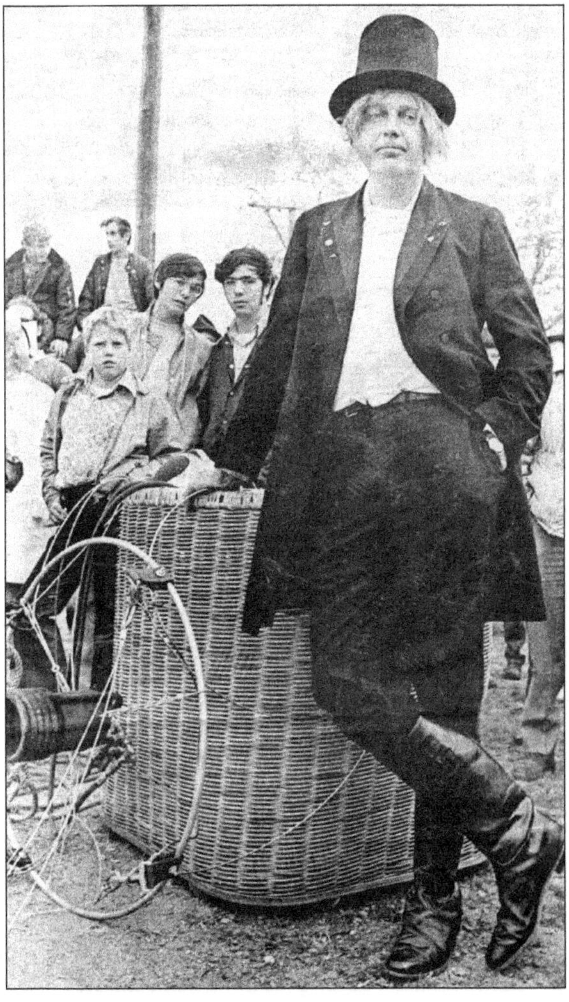

Dressed like a baloonist of the 1850s, Prof. Charles E. MacArthur of Tolland, Connecticut, stands by a wicker basket in which he flew over the Topsfield Fair Saturday evening in an 85-foot hot-air model in 1971. Balloonists prefer to ascend just after sunrise or in late afternoon, when the winds are at their calmest. These lighter-than-air craft have no steering, so the calmer air makes for safer flying conditions.

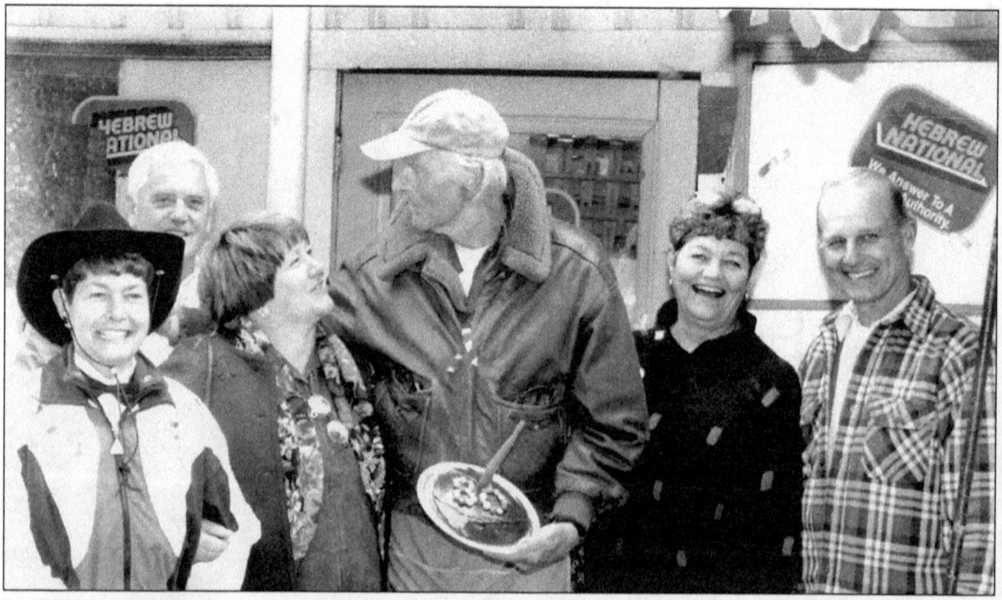

In 2001, a few friends of Larry West gave him a birthday party. Known to many fairgoers as "Larry Whistle," West was easily recognizable from his red-and-blue wig and the whistle he carried in his mouth to attract attention to his concession stand near the west entrance to the Grange building. He was the oldest concessionaire on the fair's roster when he passed away in late summer 2002. Pictured here are, from left to right, Barbara Carter, fair souvenir booth manager; Alvin Craig, manager of the fair; Andrea Derasto, fair receptionist; Larry West; Sally O'Maley, the fair's director of advertising and public relations; and Dana Carter, the concessions manager.

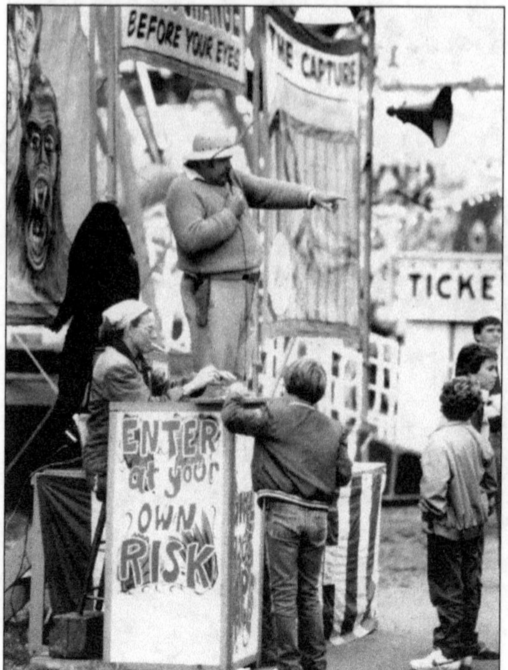

Carnivals and fairs everywhere have midway shows similar to the one shown here at the Topsfield Fair. The operators of these shows were known as "fakers."

Without cotton candy, would we really be at the fair? Cotton candy has been one of the staples of fairgoers for a very long time. From the beginning, this spun-sugar delight has always been pink, but in recent years, other colors (such as blue, green, and purple) have made an appearance. This unidentified girl at the Topsfield Fair seems to be enjoying her cotton candy no matter what color it is.

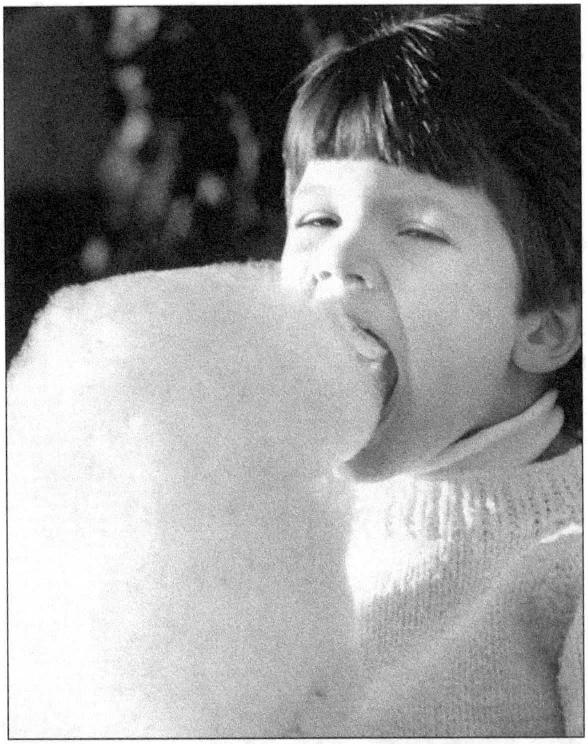

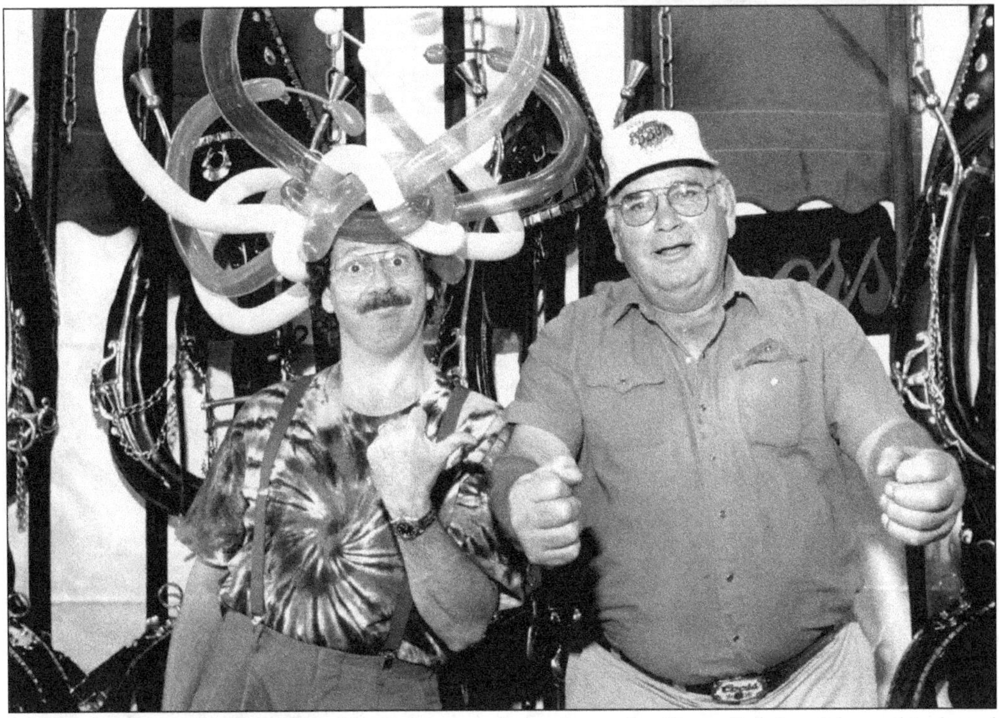

Balloon magician Lon Cerel and Dick Sparrow (who drove the Coors Belgian 6-horse hitch and the world-famous 80-horse hitch) are both in the *Guinness Book of World Records* and are both appearing at the Topsfield Fair.

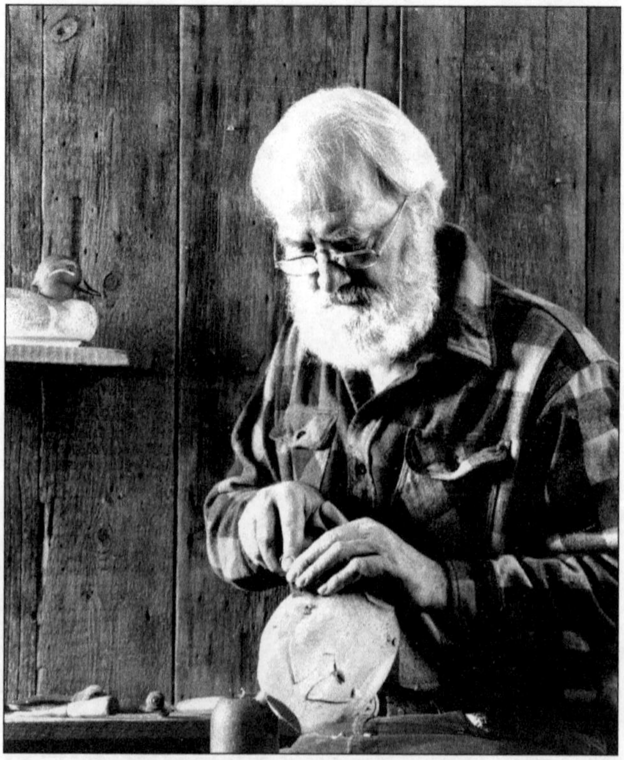

This world-renowned waterfowl decoy carver and movie actor demonstrated his carving skills for many years in the Grange building's eastern entrance. Robert Brophy of Essex for years greeted people at his booth while he carved his decoys. Brophy loved to talk wildlife and decoys to friends new and old as they wondered at his craft. One of his fine art-type decoys of a male wood duck is on a shelf behind him in this timeless photograph. He also made simpler decoys to "shoot over" as well as the fine-art type, showing great details in feathers and colors beyond belief. They are so realistic that they look like taxidermy mounts. Brophy recently played the part of the constable in a film about the Salem witch trials, *The Crucible*, filmed in Ipswich and starring Winona Ryder.

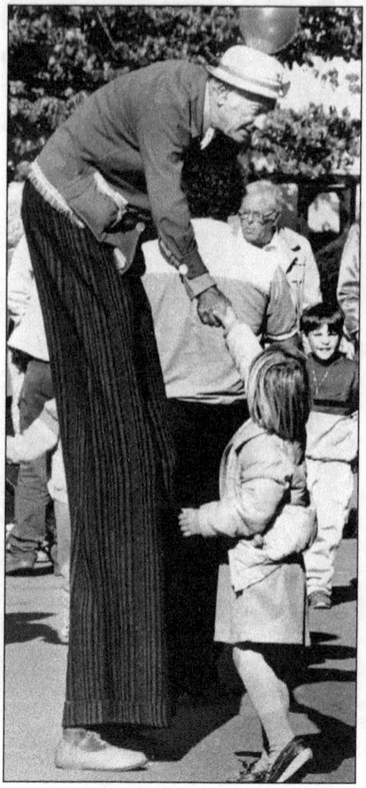

A tall man who was a perennial favorite of children and adults is Mike Monroe of Wakefield. "Mr. Tall Top," as he was called, precariously bends down on his stilts to greet Donette Pelletier of Newburyport at the Topsfield Fair in 1986.

In the earlier fair days, vaudeville acts were very popular, and as the century came to a close, country western performers became popular when they came to the fair to sing on the stage between the poultry and fruit-vegetable buildings. Crystal Gale was among the most popular performers to sing in the free admission shows at the fair in 1995.

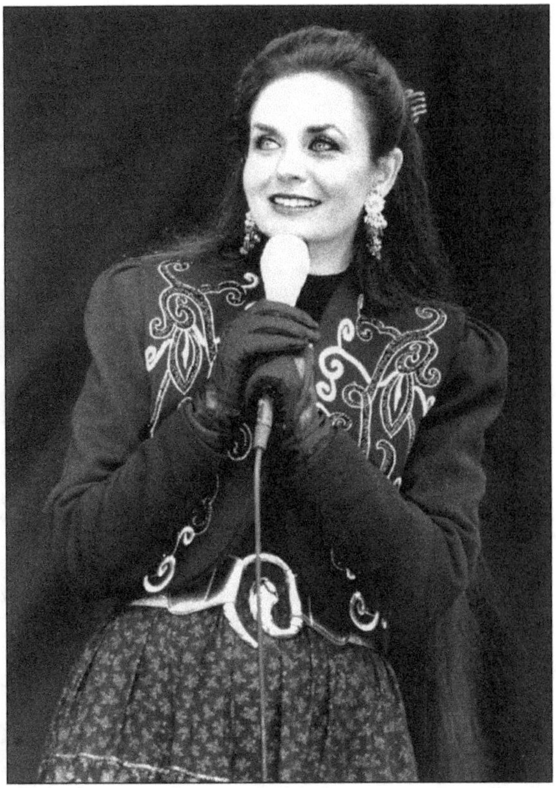

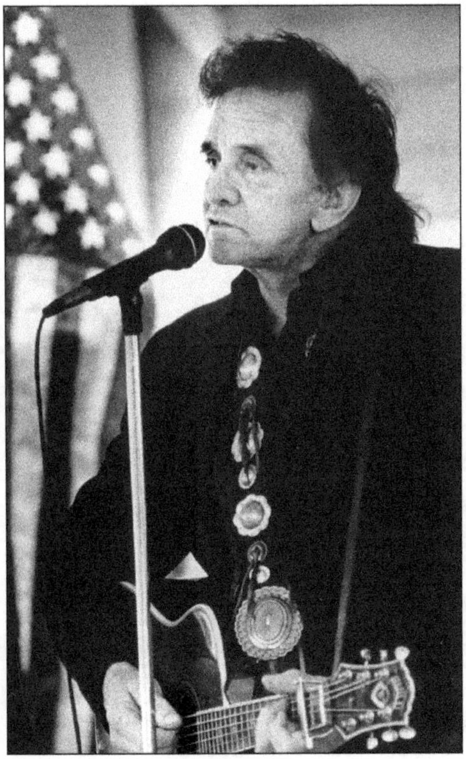

Johnny Cash was another country western singing star who always impressed the crowd when he performed on the stage near the center oval.

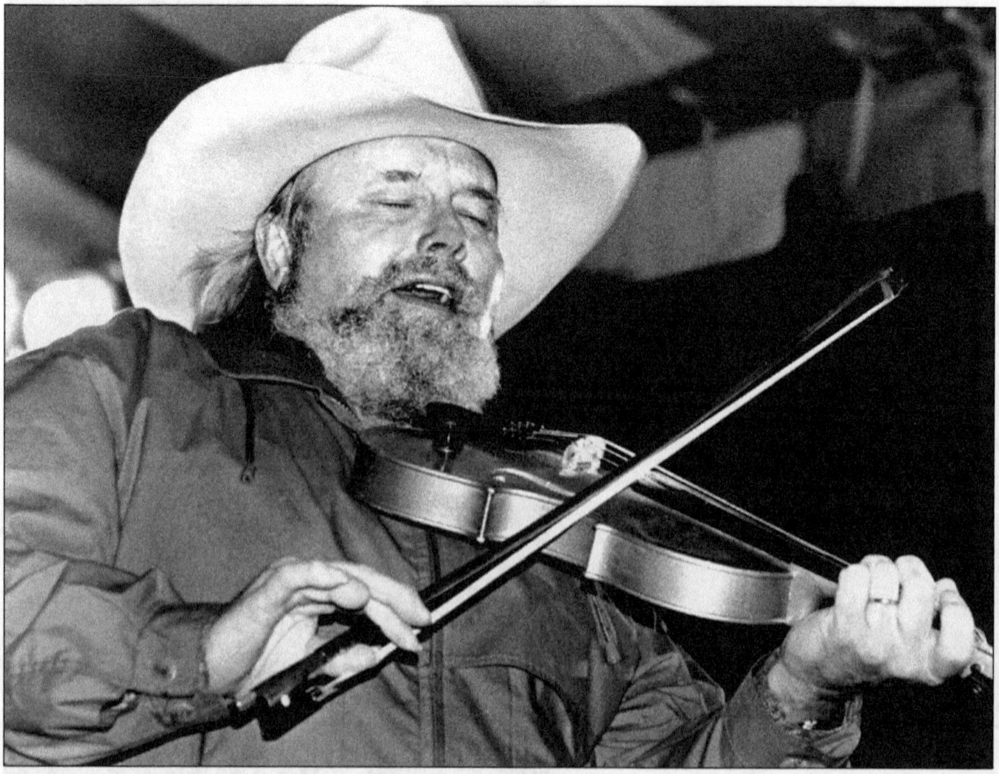

Country western performers were very popular when they performed at the fair. Charlie Daniels, with his prowess on the fiddle, was no exception.

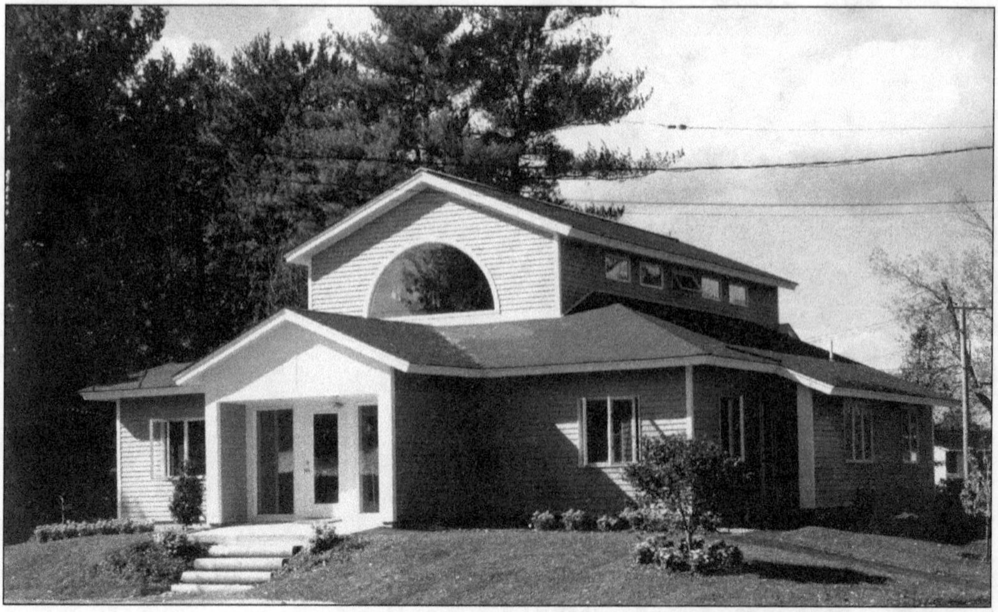

After years of cramped quarters in the old administration building, the Essex Agricultural Society, which stages the Topsfield Fair every fall, finally had a new spacious and attractive administration building constructed on the site of the previous building in 1986.

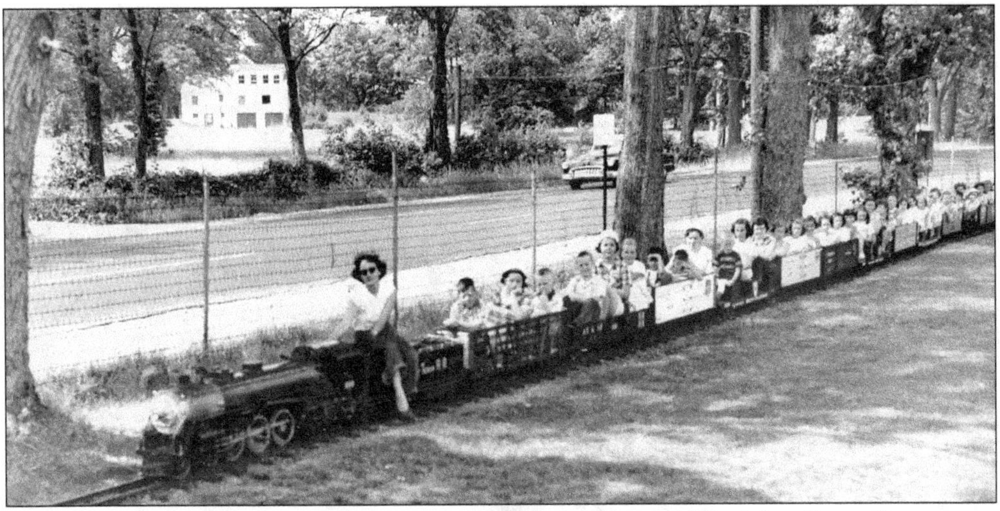

Along Route 1 (the former Newburyport Turnpike), the Topsfield Fair's Joy Town 18-car train, with 2 cabooses, passes in the early 1950s with a load of 54 passengers, including both children and adults. This locomotive, weighing 550 pounds, is a scale model of a Boston and Albany Railroad No. 350 Hudson 4-6-4. The train operates on the permanent $7\frac{1}{2}$-inch-gauge track in the fair's park. The engine now belongs to the Masons at their retirement home in Charlton. (Photograph courtesy of the Beverly Historical Society, the Walker Collection.)

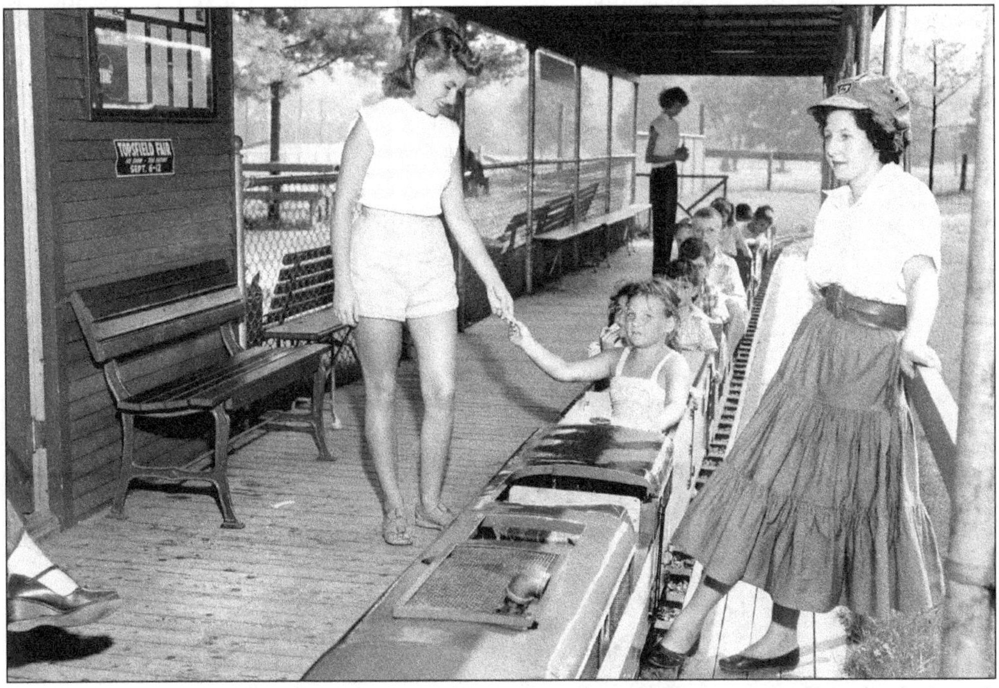

The Joy Town diesel train was an extremely popular ride with the youngsters in the fair's park area next to Route 1 during the 1950s and later. Here we see Miss Topsfield Fair during 1951, Marilyn "Bunny" Atkinson, helping the train's engineer, Harriet Heard of Ipswich, by collecting ride tickets from patrons on the train. Harriet later married Rusty Stanwood of Topsfield. The Stanwoods, all relatives of the author, were early tenant farmers at the society's Treadwell Farm in the 1890s.

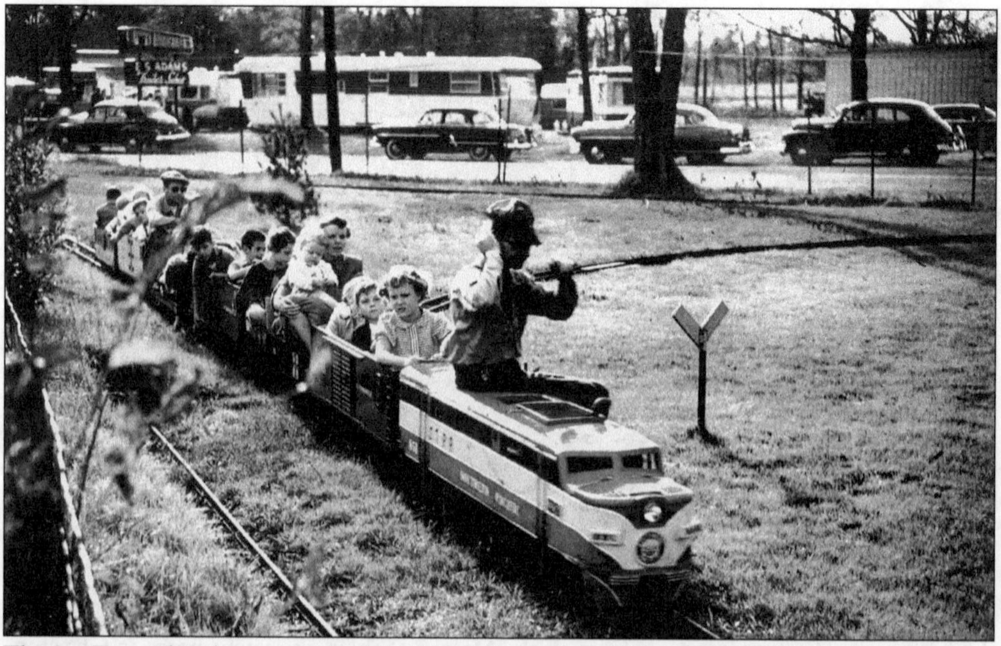

The Joy Town diesel train was built by Lester D. Friend of Danvers *c.* 1950. In this 1955 view, the train is transporting children around the Joy Town Kiddie Park near Route 1 behind the fair's administration building. (Photograph courtesy of the Beverly Historical Society, the Walker Collection.)

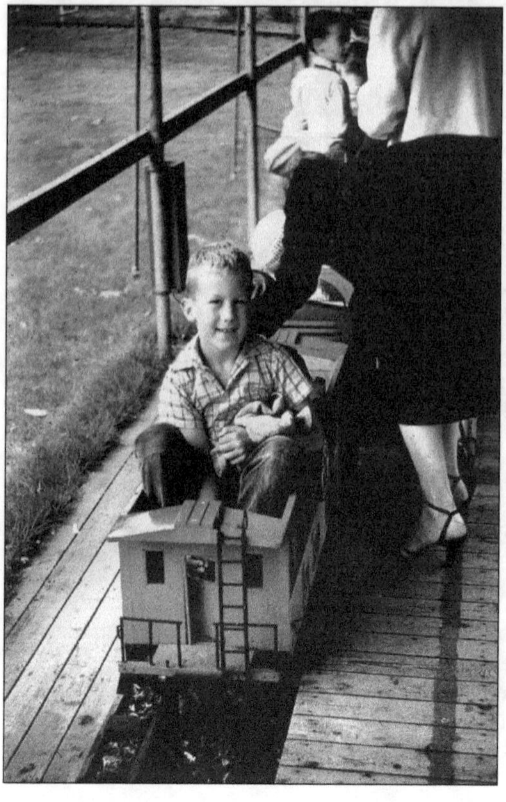

In 1955, Chip occupies the very best seat in any train, the caboose. This Joy Town train was built by Lester D. Friend of Danvers. In 1955, and for many years thereafter, thousands of youngsters rode the diminutive train through the park at the fair. (Photograph courtesy of the Beverly Historical Society, the Walker Collection.)

The Miss Topsfield Fair of 1953, Marilyn "Bunny" Atkinson of Ipswich, is shown helping to control the Joy Town roller coaster ride by holding the large brake lever. Thousands of children over the years were delighted to ride this contraption and have a thrilling time.

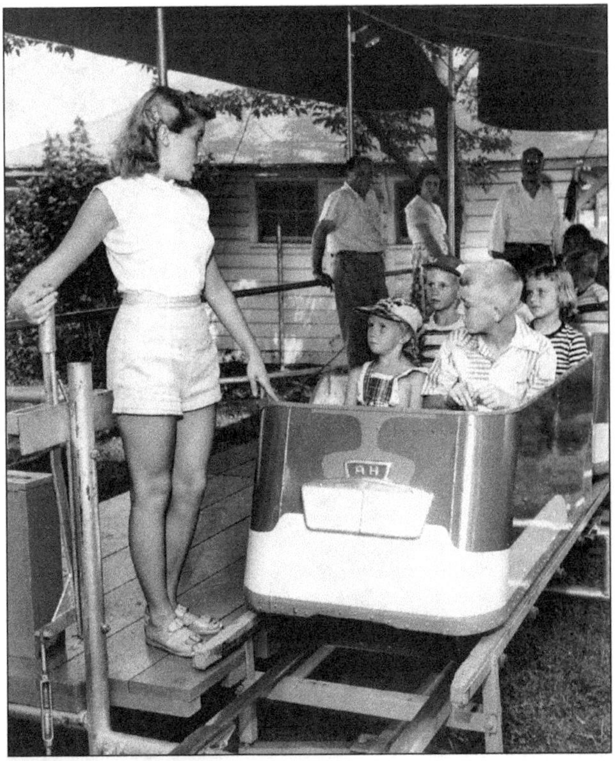

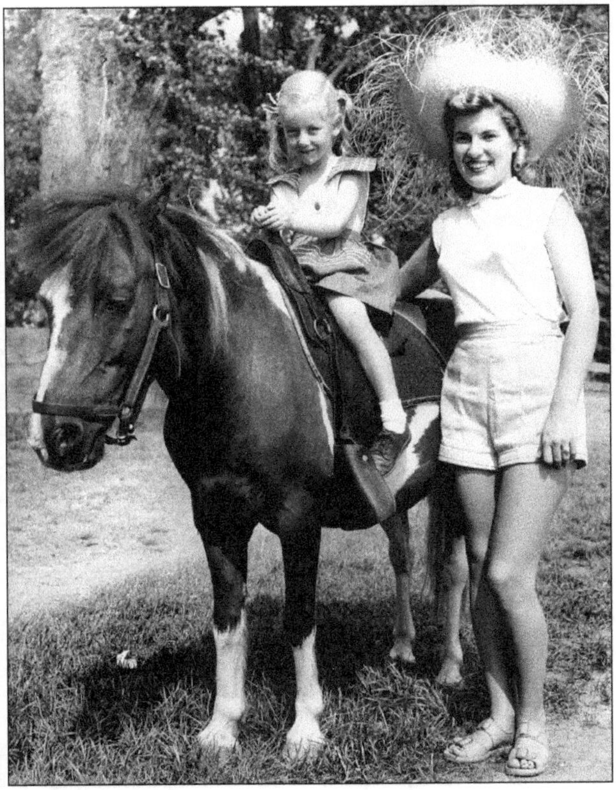

Marilyn Atkinson helps Christine Leonard of Beverly on the pony rides at the fair in 1953.

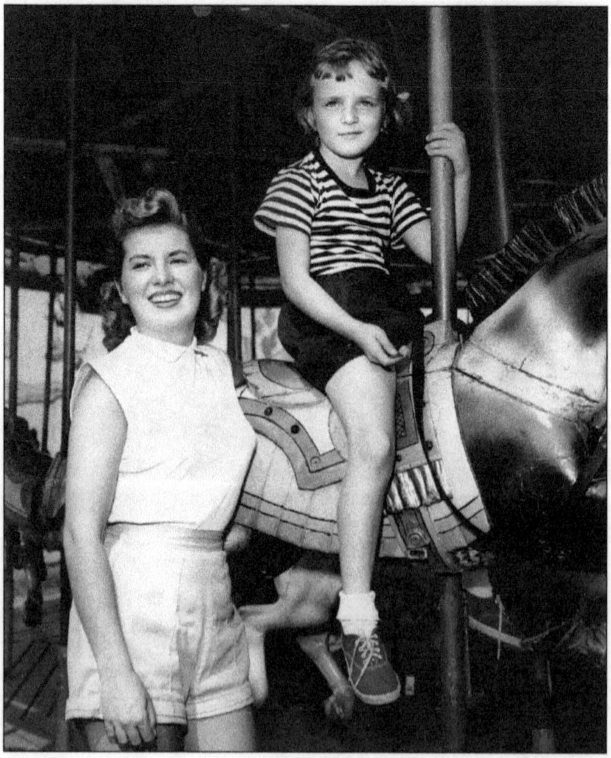

Marilyn Atkinson accompanies six-year-old Kate Martin of Middleton on the flying horses at the 1953 Topsfield Fair. The carousel was part of the Joy Town rides.

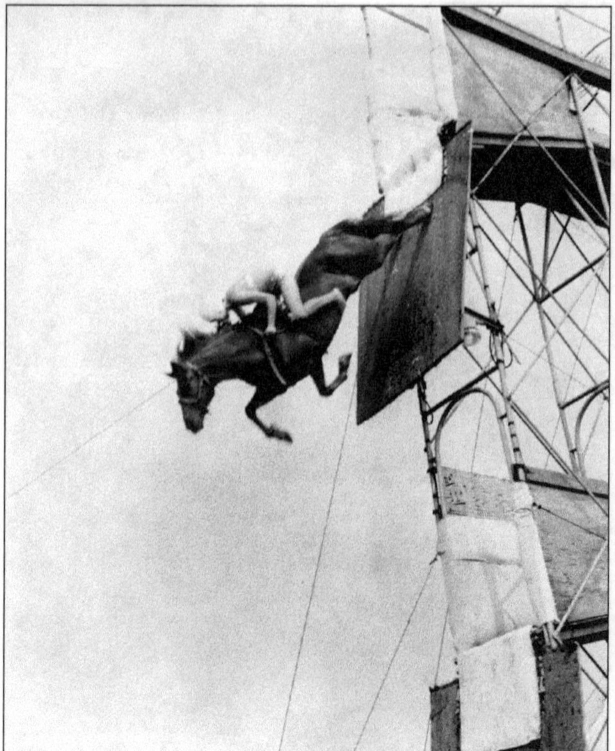

Shown is a superb photograph of Dr. Carver's famous diving horse and rider in 1959. The tower was nearly 60 feet tall, and the tank of water was quite small. At the Topsfield Fair over the decades, both men and women dove into similar pools of water on both horses and motorcycles. Sometimes men were known to have missed the pool, as one did at a previous event at Topsfield, narrowly escaping with his life. (Photograph courtesy of Barbara and Daniel Hurld.)

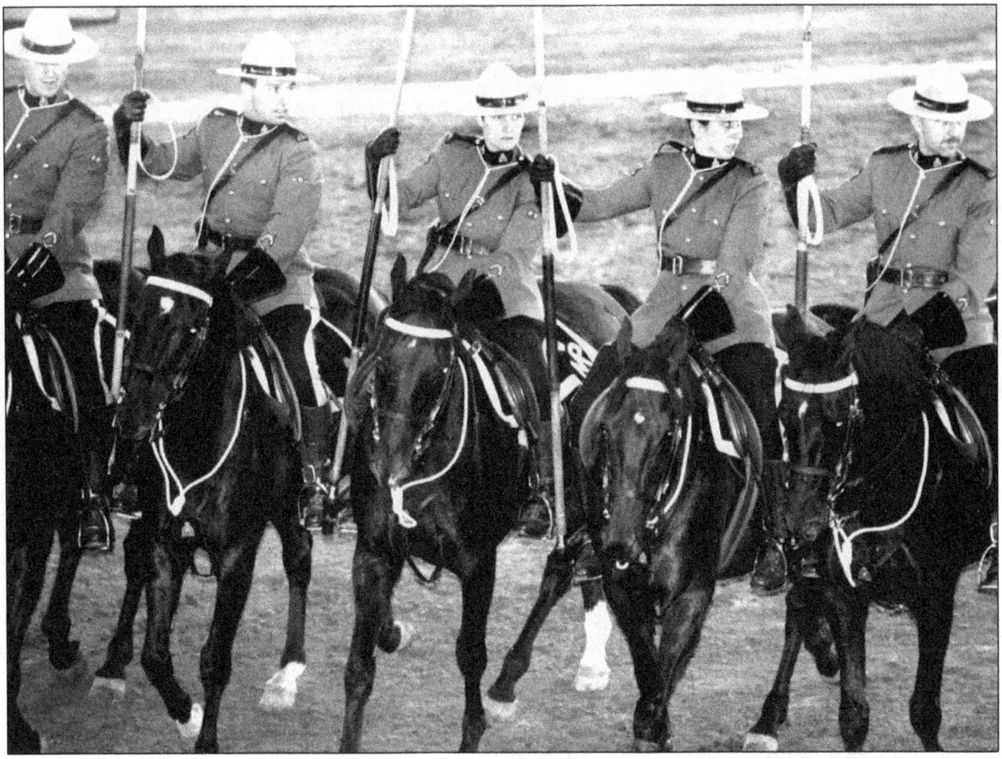

When the Royal Canadian Mounted Police performed at the fair in 2000, they executed the star maneuver. Here we can see two women in the ranks of the Musical Ride. The mounted police first came to perform at the fair in 1970 and now come about every two years.

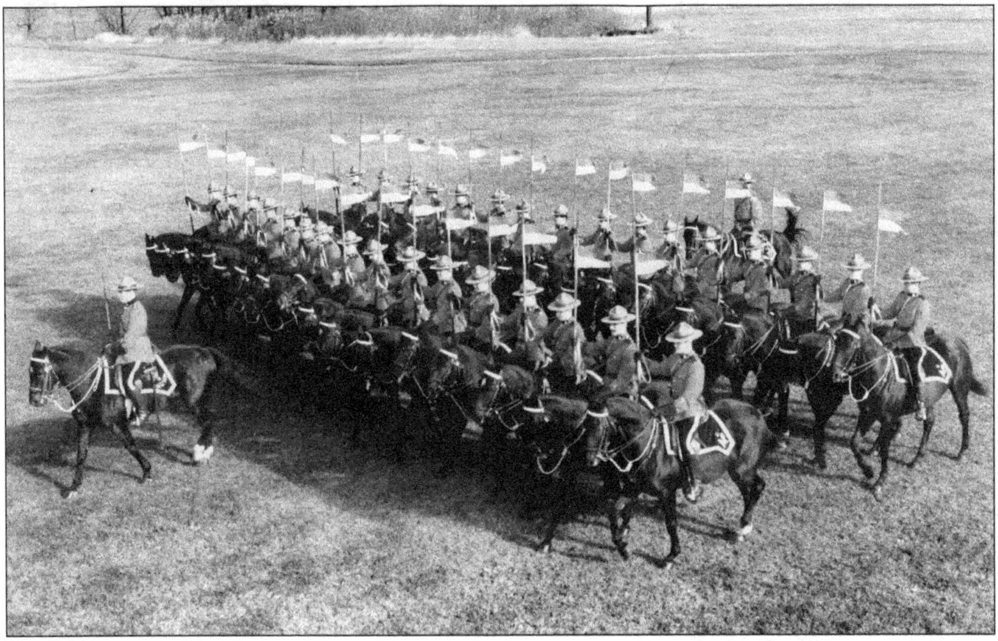

The mounted police perform their world-famous Musical Ride daily when they come to the Topsfield Fair. In this view, a troop is drilling on the practice field.

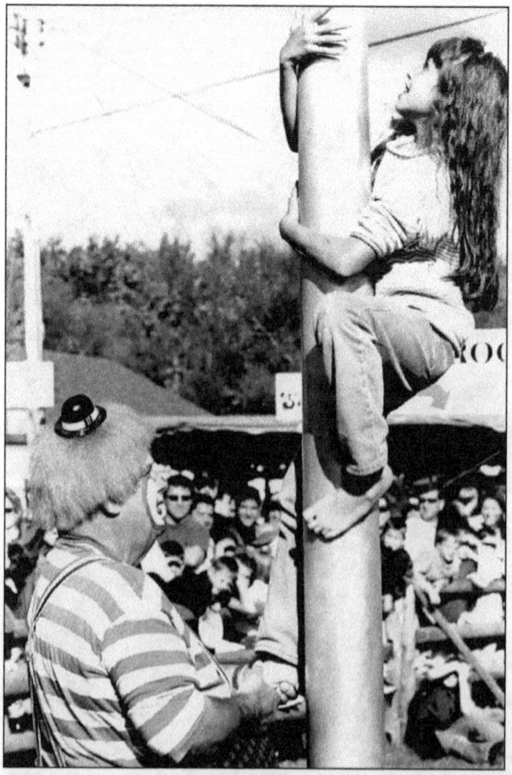

Taylor Awnese of Pepperell climbs the Money Pole in the Old Essex County Barnyard in 2002 while Dune Buggy the Clown lends a helping hand. The Money Pole Climb was started in the 1970s by Daniel Hurld, host of the Old Essex County Barnyard for many years.

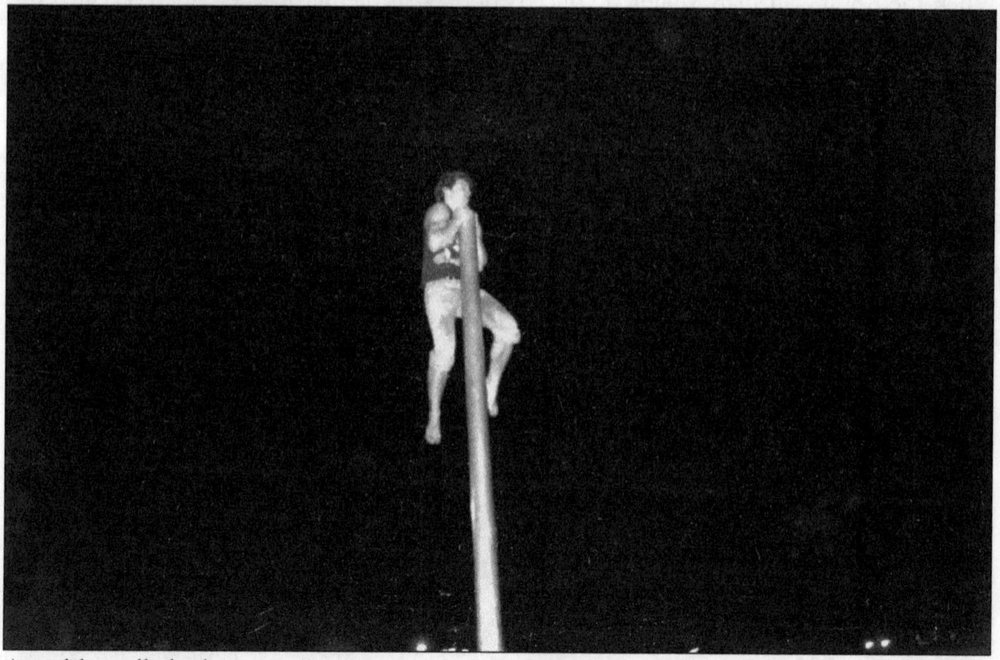

An athletically built young man is successful in climbing easily to the top of the Money Pole at night to collect his $10 prize money in the 1990s. Night success on the pole was rather rare as the aluminum tapered pole gets cold at night and develops a damp surface that makes it extremely difficult to grip and to climb.

"Double take" is the only phrase to describe the actions of Ipswich twins Matthew and Austin Sharpe. The 11-year-old boys both competed in the Money Pole contest, and each reached the top to win a ribbon and $10 prize money at the 166th annual Topsfield Fair in 1984. Daniel Hurld, the host at the Old Essex County Barnyard, thought the brothers were pulling a fast one on him by competing twice until he spotted the first climber was a twin. The boys returned in subsequent years to repeat their first success.

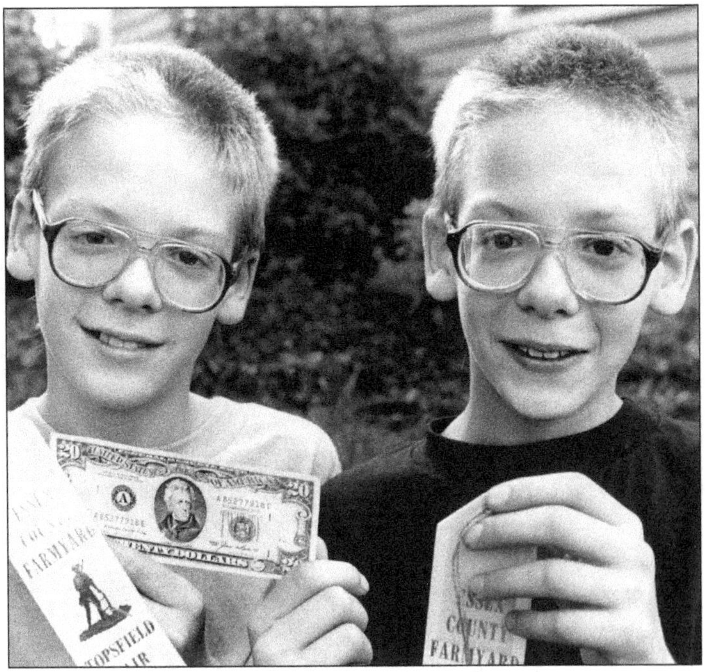

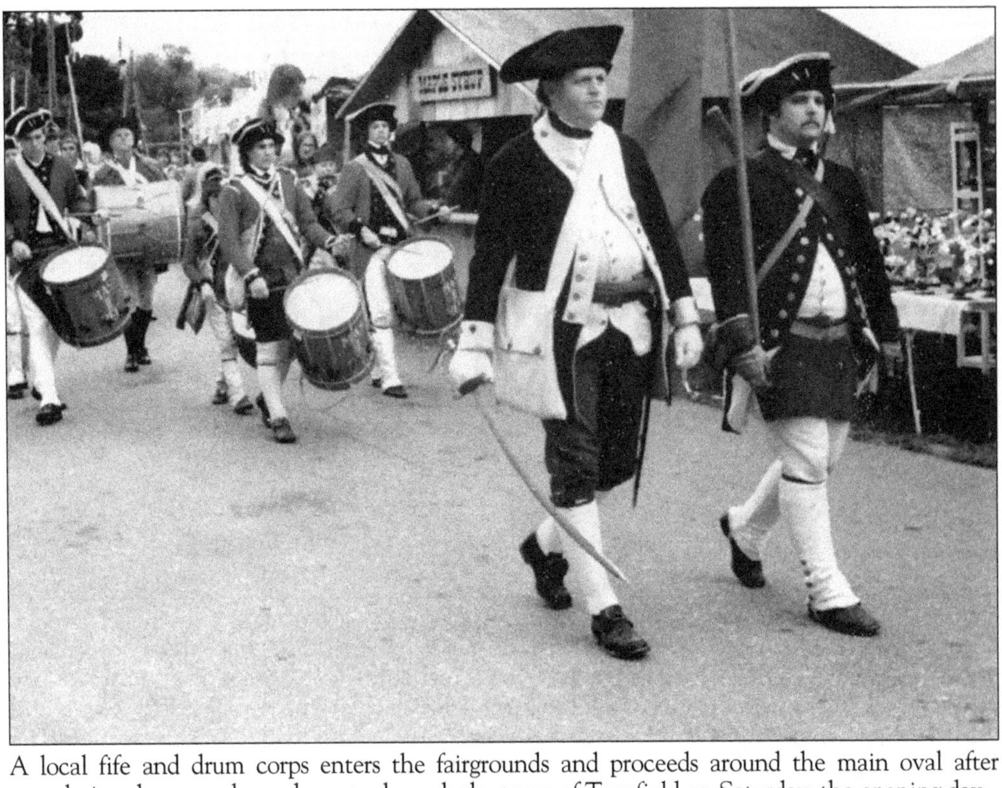

A local fife and drum corps enters the fairgrounds and proceeds around the main oval after completing the annual parade route through the town of Topsfield on Saturday, the opening day.

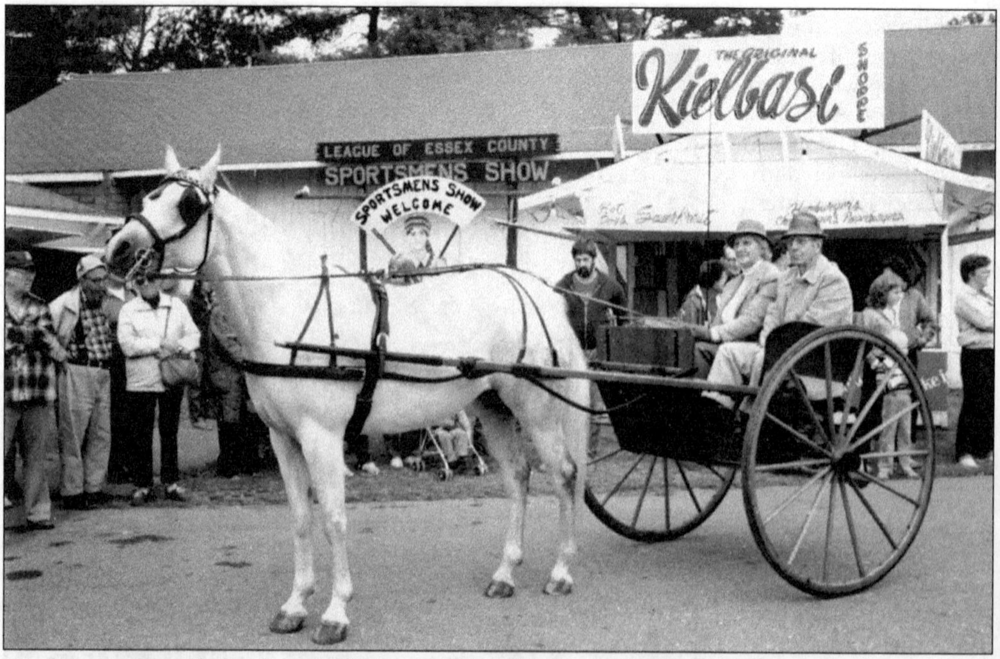

At the opening of the Topsfield Fair in 1982, Barbara Hurld of West Newbury drives her Arabian mare to a road cart as her father, Donald Shepard, enjoys himself along the parade route ending on the fairgrounds. The building in the background was originally built as a dance pavilion. It was later used for the 4-H exhibits and now serves as the sportsmen's show building. (Photograph courtesy of David H. Fletcher.)

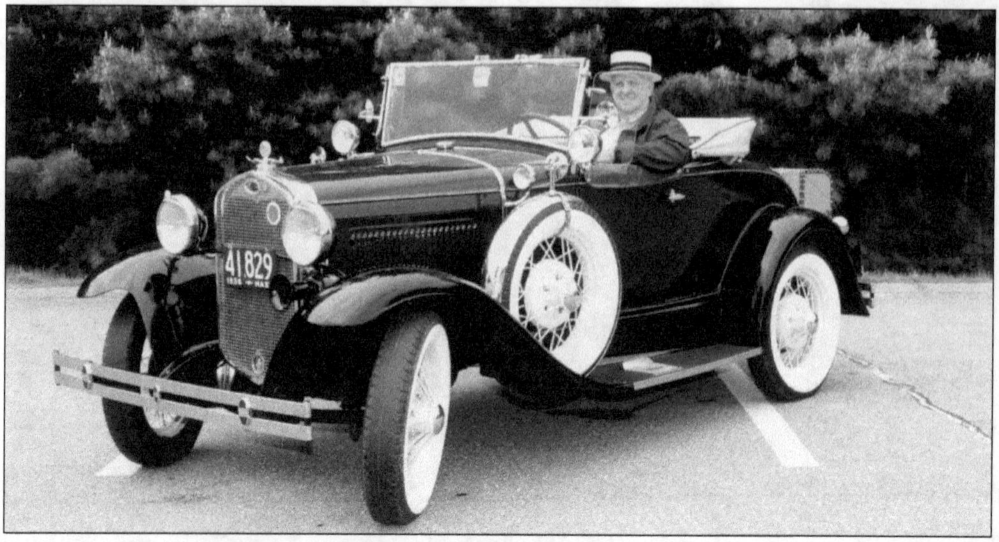

Each year for the opening of the fair, a parade is held along the main streets of Topsfield. Beginning at the town common and ending at the fairgrounds, the parade winds its way through the crowds and out into the main parking lot. This photograph shows one of the antique automobiles that has participated in the last several fair parades. The author awaits his place in the parade procession in his restored, all-original-stock 1930 Ford Model A open roadster. The car has its original 40-horsepower four-cylinder engine and transmission. (Photograph courtesy of David H. Fletcher.)

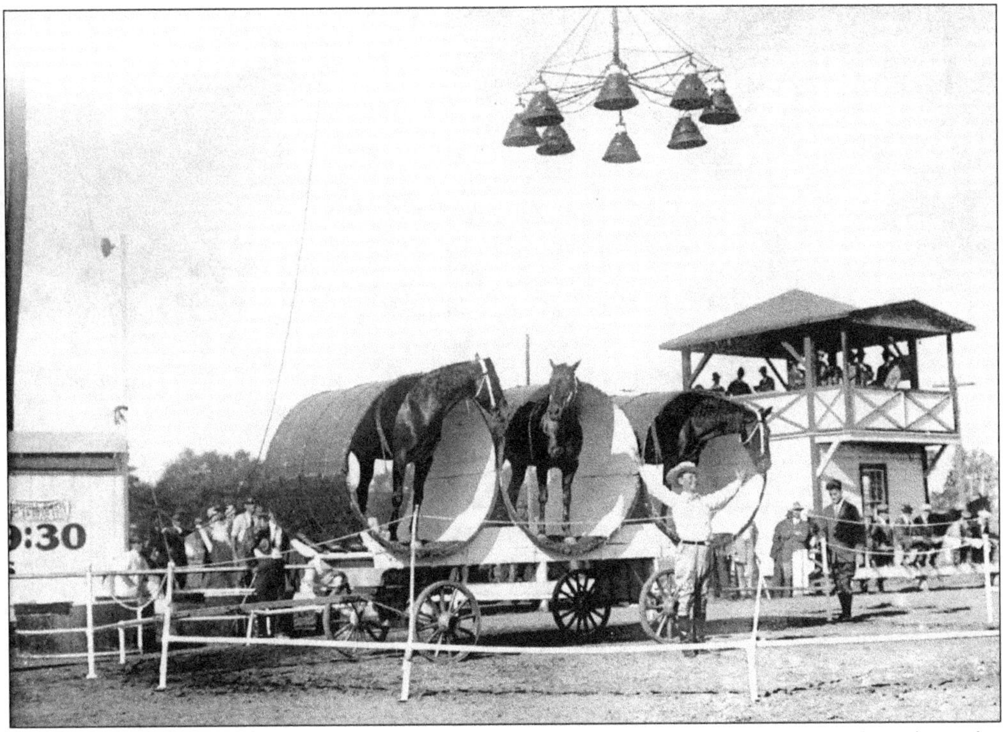

The Agee's Trained Horse act, with the trainer-performer, has three horses in barrels at the grandstand show at the Topsfield Fair in 1927.

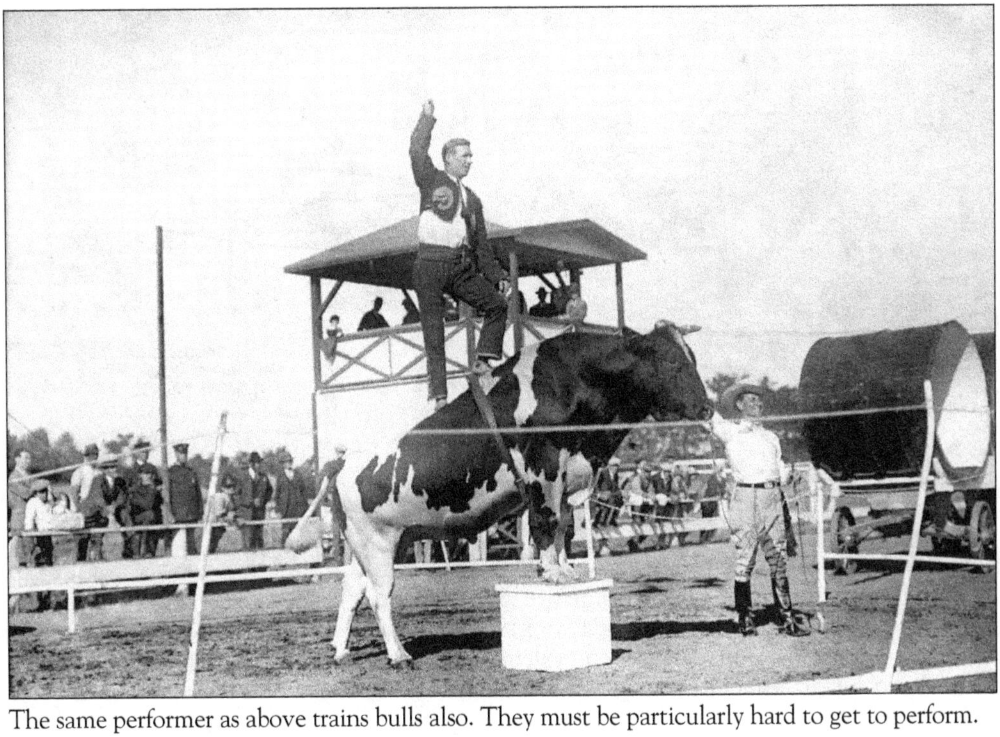

The same performer as above trains bulls also. They must be particularly hard to get to perform.

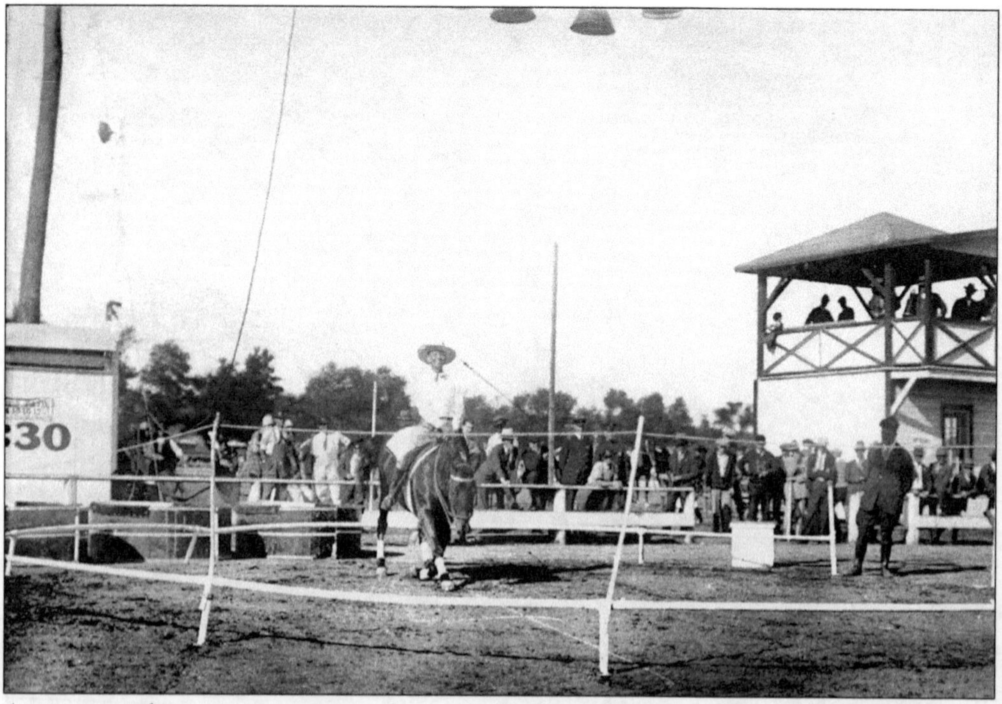

Again we see the trainer-equestrian riding one of his fine show horses and ending his performance with a bow in 1927. These animals were part of the Agee's Trained Horse act.

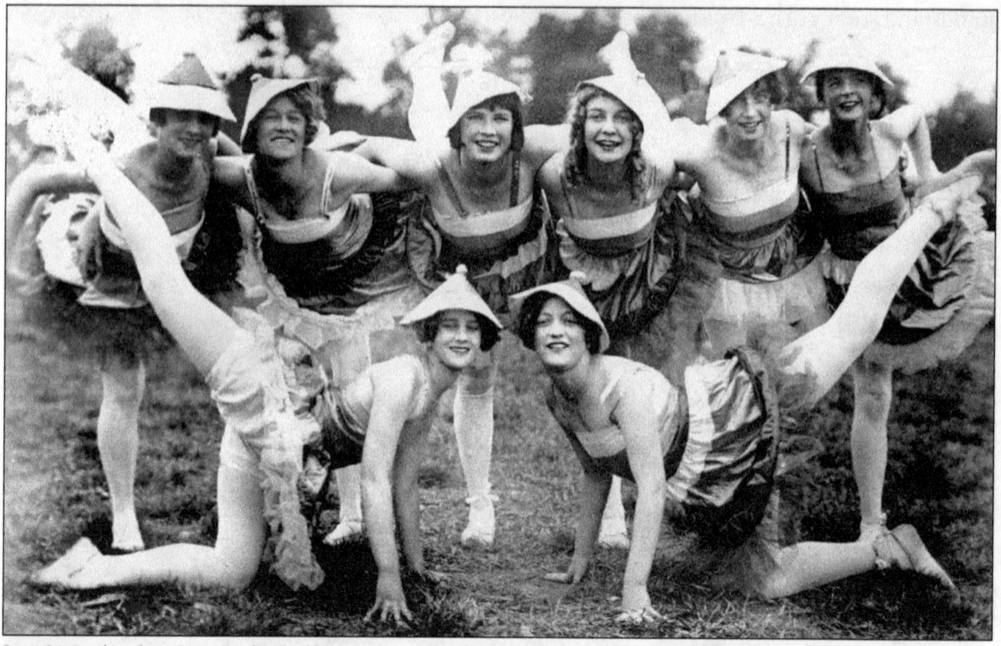

In 1928, the fair featured a ballet performance with these eight ballet dancers performing on the grandstand area stage. During the period from 1910 to the start of World War II, vaudeville acts and other thrilling performances were prevalent at the fair. Many special animal and high-wire acts were booked to attract large crowds besides the thrill rides on horses, motorcycles, bicycles, and automobiles

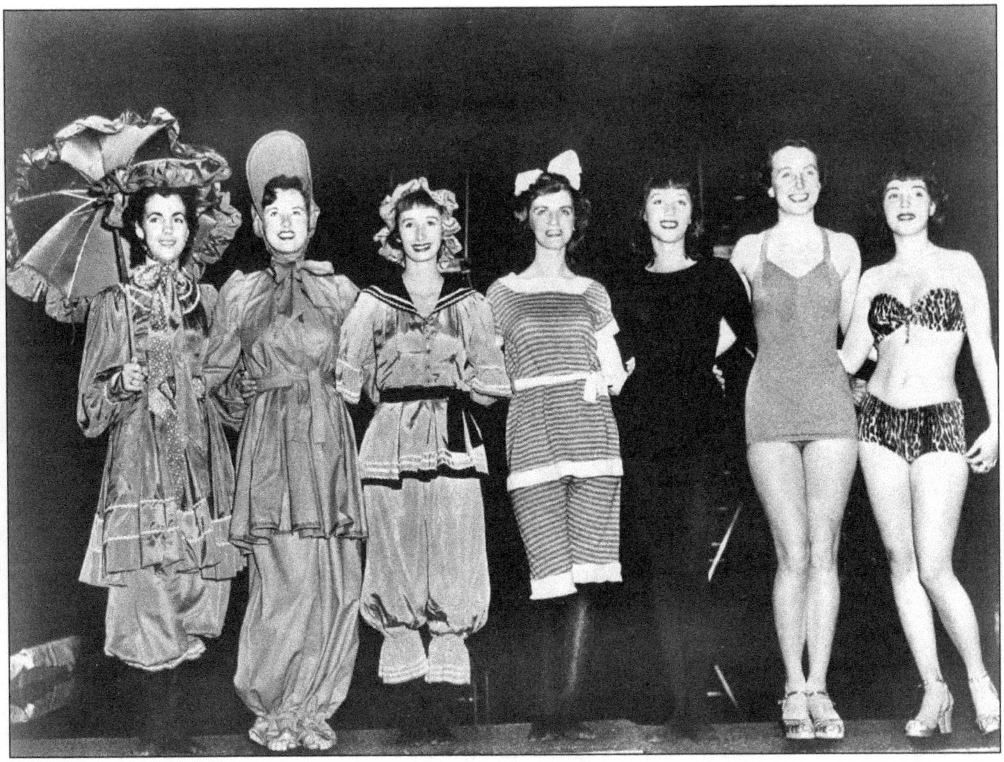

The Water Follies shows of the early 1950s were a crowd pleaser. As part of the 1951 follies, the lady swimmers from the show's cast put on a bathing suit exhibition, with the fashions ranging from 1850s to 1950s styles.

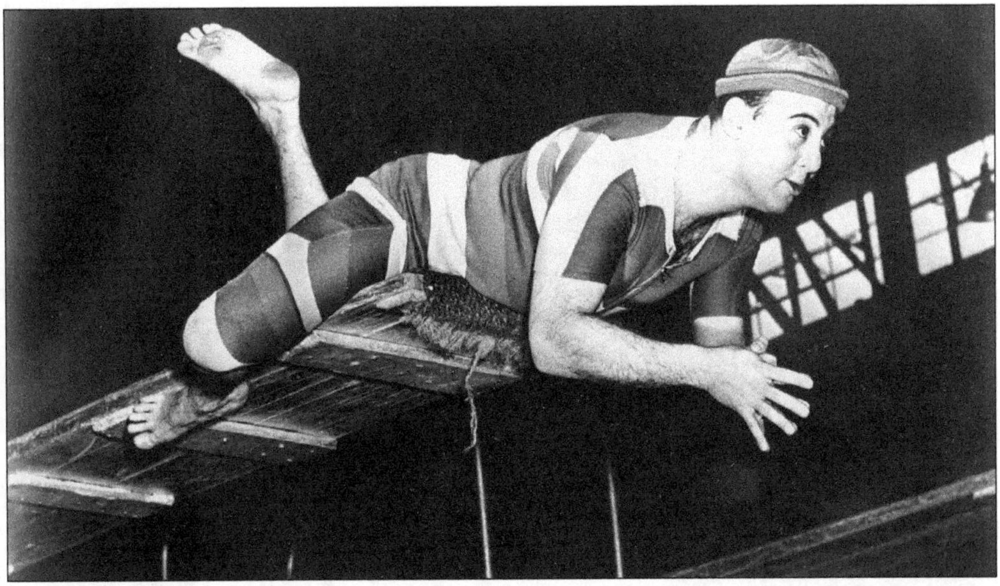

Eddie Rose, diving's "Clown Prince," provided laughs with his popular comedy diving antics at the Water Follies presented twice daily, from August 31 to September 6, 1952, at the Topsfield Fair. The Water Follies were one of the major entertainment highlights of that annual agricultural fair.

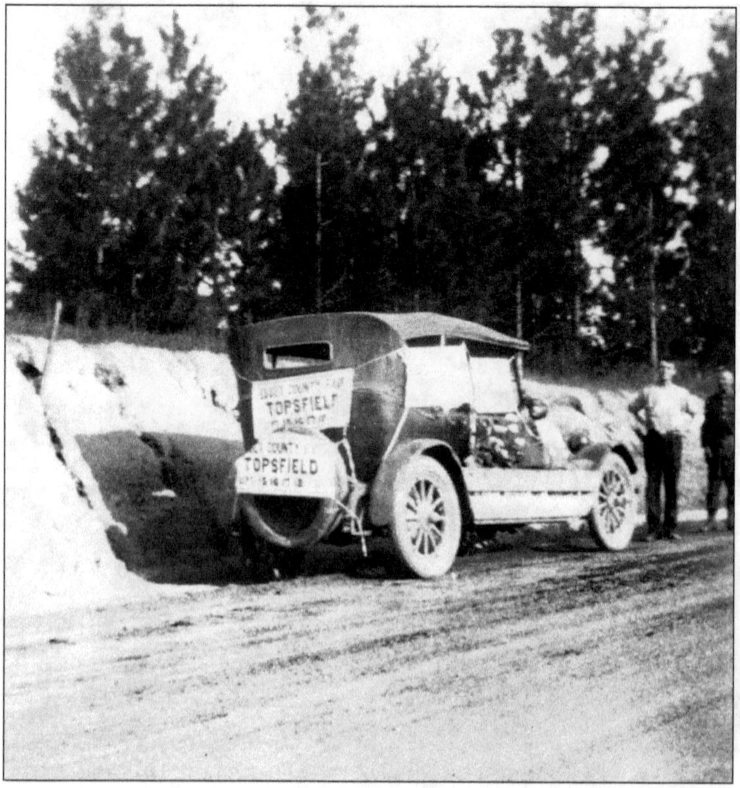

In 1926, a touring car is decorated with banners proclaiming the dates for visiting the Essex County Fair. The banners were used on automobiles the same way the advertising tire cover that is now at the fair office was used—to create a traveling billboard.

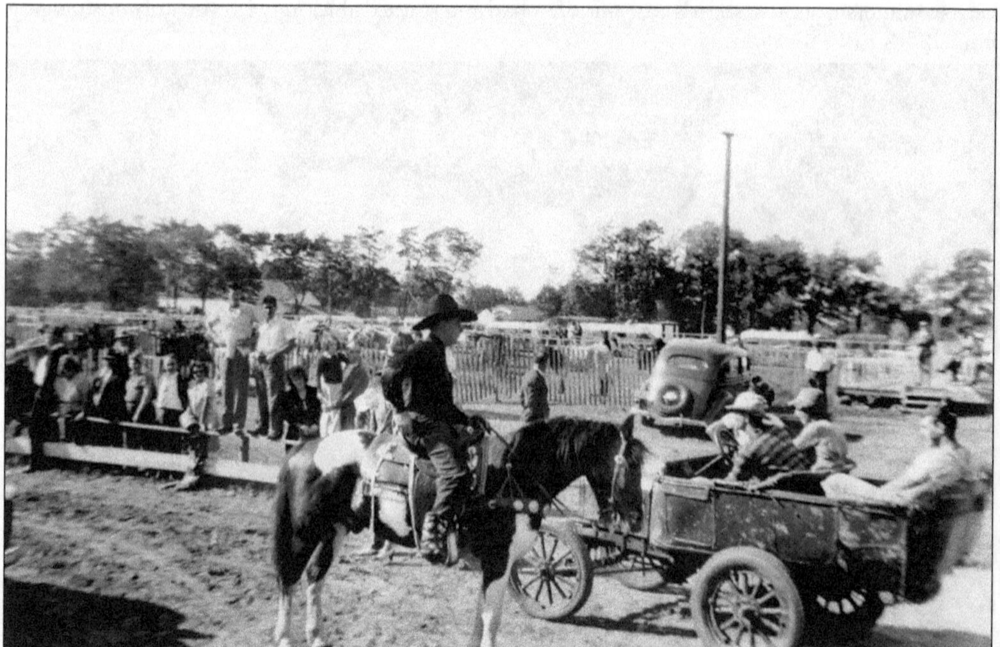

The fair presented rodeos over the years. In 1941, Hoot Gibson appeared at the Topsfield Fair to take part in the Western-style show. We also can see the clown car that is still around today at shows in New England. (Photograph courtesy of Yolanda Marciano.)

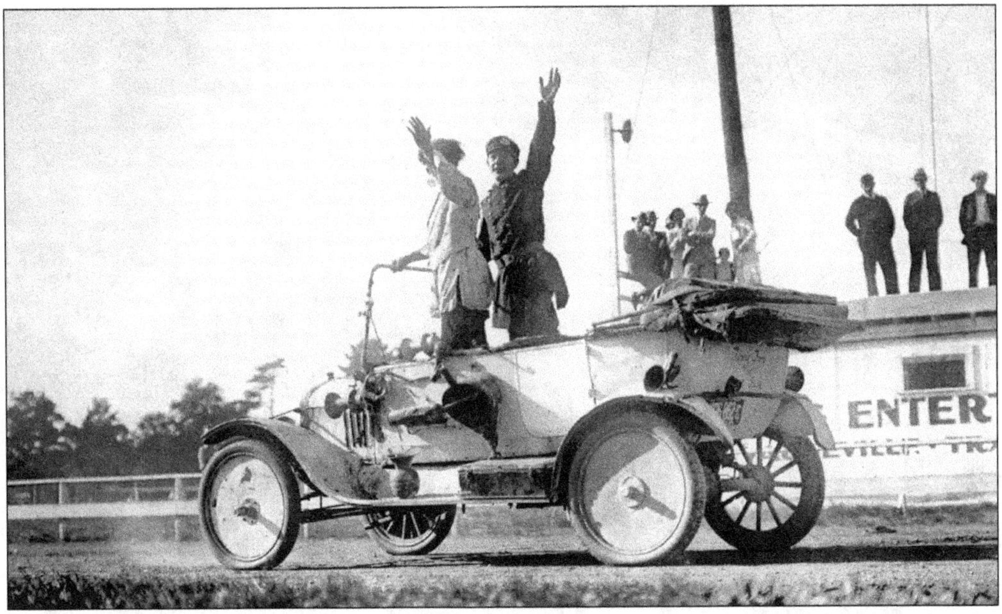

Clowns were always part of circuses and carnivals, and fairs were no exceptions. They played a big part in vaudeville, and vaudeville played a big part in the entertainment of fair patrons. Two of these clowns *c.* 1926 are riding in the clown car in front of the vaudeville stage, with its high-wire apparatus, across from the grandstands at the fair's racetrack. The old car has both disc and wooden spoked wheels and even has a spittoon on the running board.

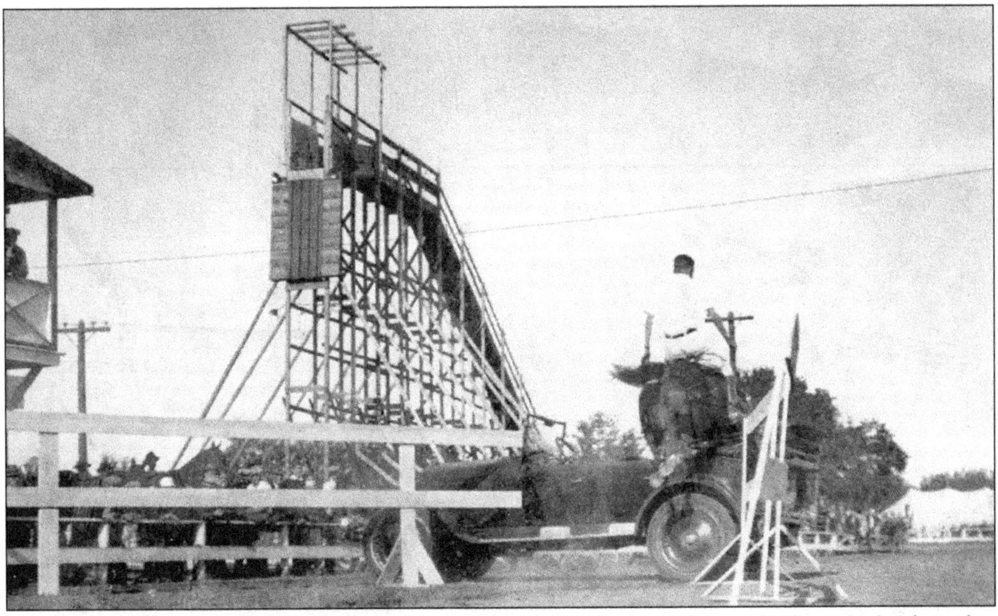

One of the most spectacular events at the Essex County Fair (as it was called in 1926, when this photograph was taken) was the horse that dove from this 60-foot platform into a large tank of water. This act was performed every day and was popular at the fair for decades after 1926, even when motorcycles replaced the horses to do the diving. Men and women on both horses and motorcycles performed these dives concurrently on their show circuit. Some of the riders were known to have missed the tank, but the horses never did.

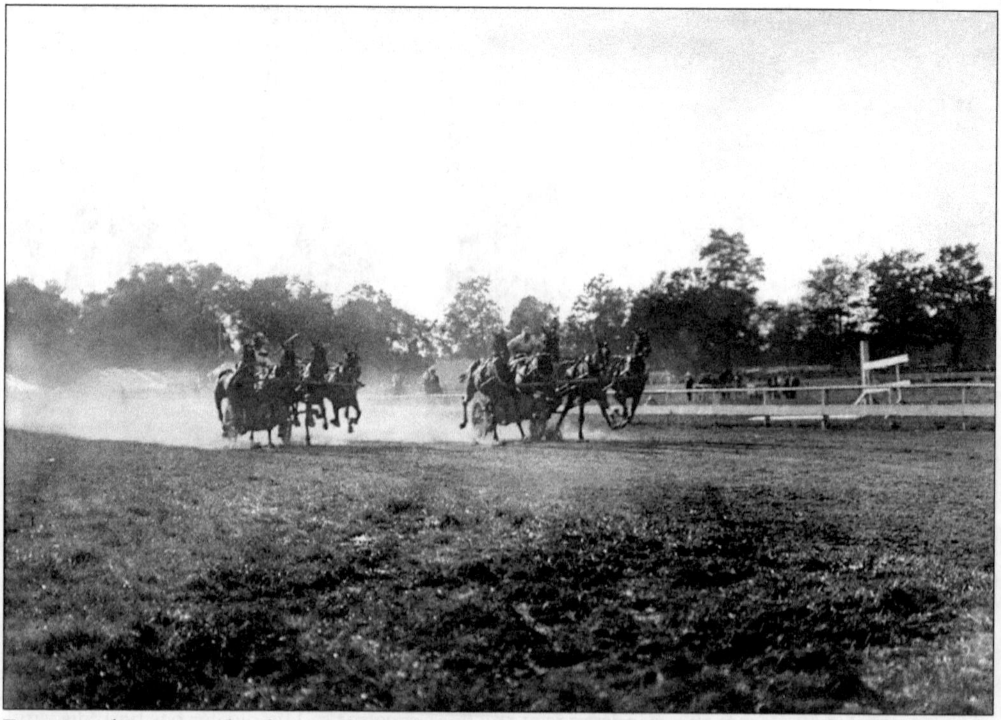

For several years at the fair in the mid-1920s, spectacular chariot races were held in which drivers of paired horses raced at top speeds around the track or even across the turf, as seen in this vintage photograph.

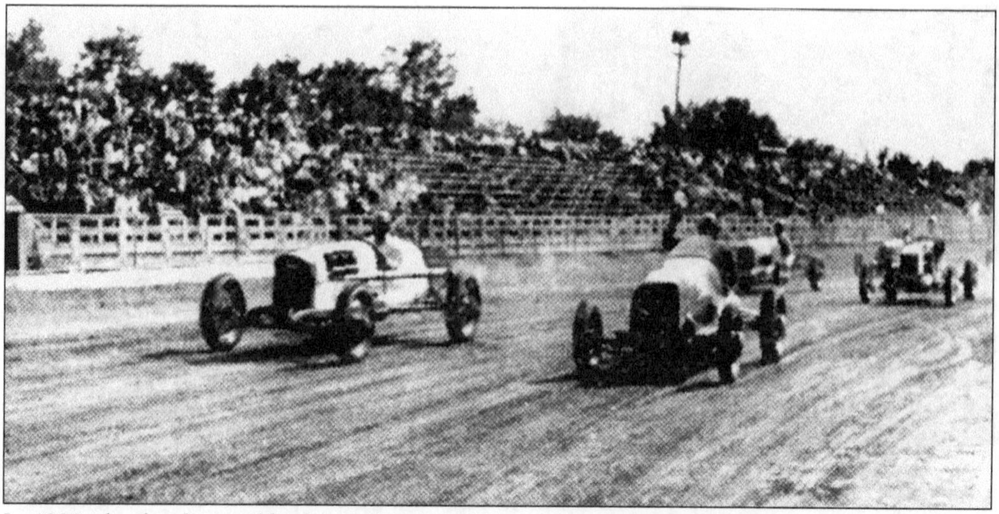

In 1938, the fair featured high-speed automobile racing, as shown by this photograph found in the 1938 fair program. This photograph was taken at a previous racing meet at an unidentified location. In a 1938 aerial photograph of the racetrack, we can see improvements being made to accommodate the high-speed racecars to be used on it.

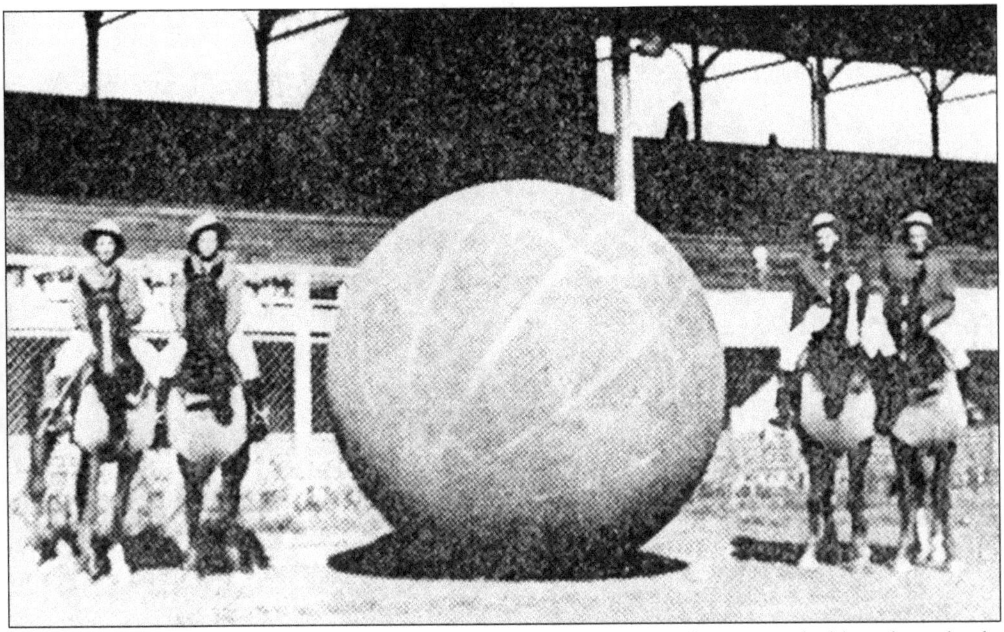

One of the highlights of the 1937 fair was the unusual Push Ball contests held on horseback. This publicity photograph was taken at another fair, but we can see the huge size of this ball. It must be nearly 10 or 12 feet in diameter judging from the horse and rider's height next to it. There is no indication in the program of the rules of the game or how the ball was even pushed, but it looks as if it would be interesting to watch. It is billed as "the largest ball in the world."

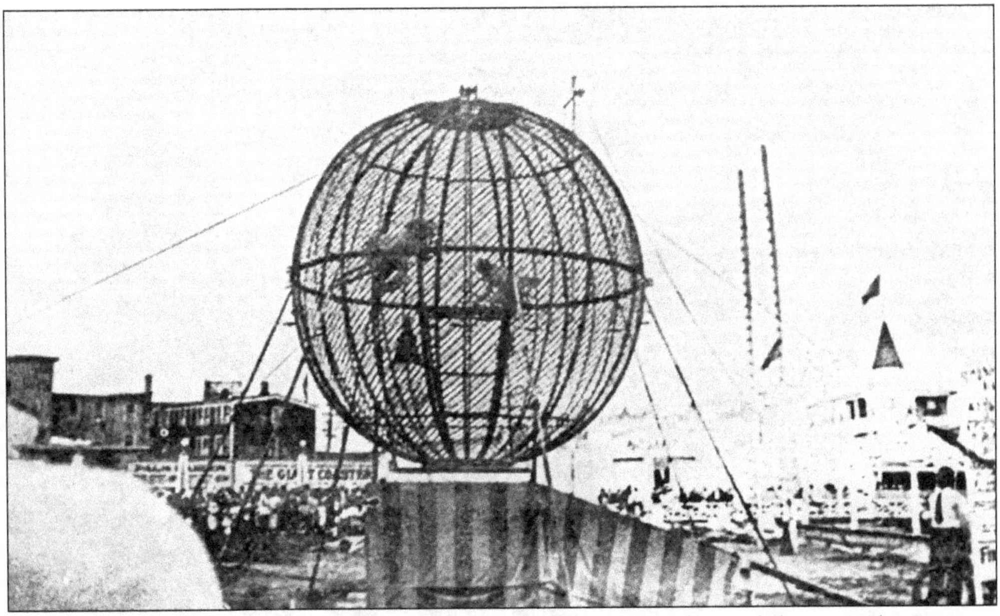

Many of us have seen motorcyclists whizzing around in a steel sphere while narrowly missing a man standing inside at the bottom of the cage, but this was a bicycle act with women doing the pedaling and a third woman standing on the cage floor below. It was no time to get a leg cramp or to run out of energy. This photograph appears in the 1938 fair show program to advertise the upcoming performances in Topsfield and was obviously taken at some other location.

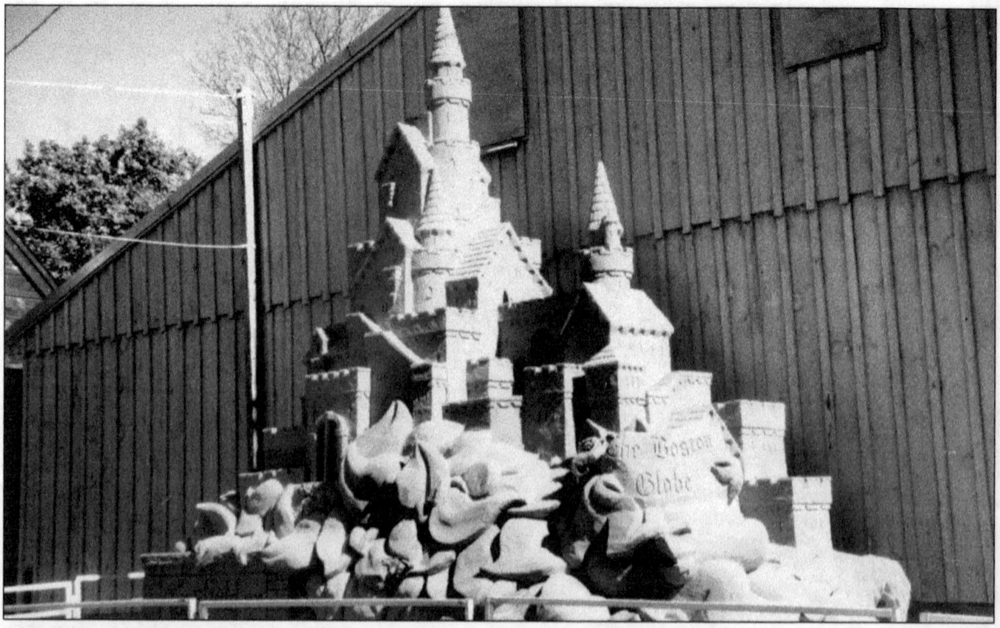

Each year, master sand castle carver Justin Gordon of Groveland builds a gigantic carved sand castle out of tons of damp sand piled high and compacted the week before the fair opens. His masterpiece is started a few days before the fair opens, so some parts are done when the patrons arrive. For several days thereafter, Gordon continues carving in the sand to create a true masterpiece. As the newspaper being read by the pig denotes, the *Boston Globe* sponsors this exhibit and many other parts of the fair. These sand castles may be seen each year between the 4-H building, the bee building, and the sheep barn.

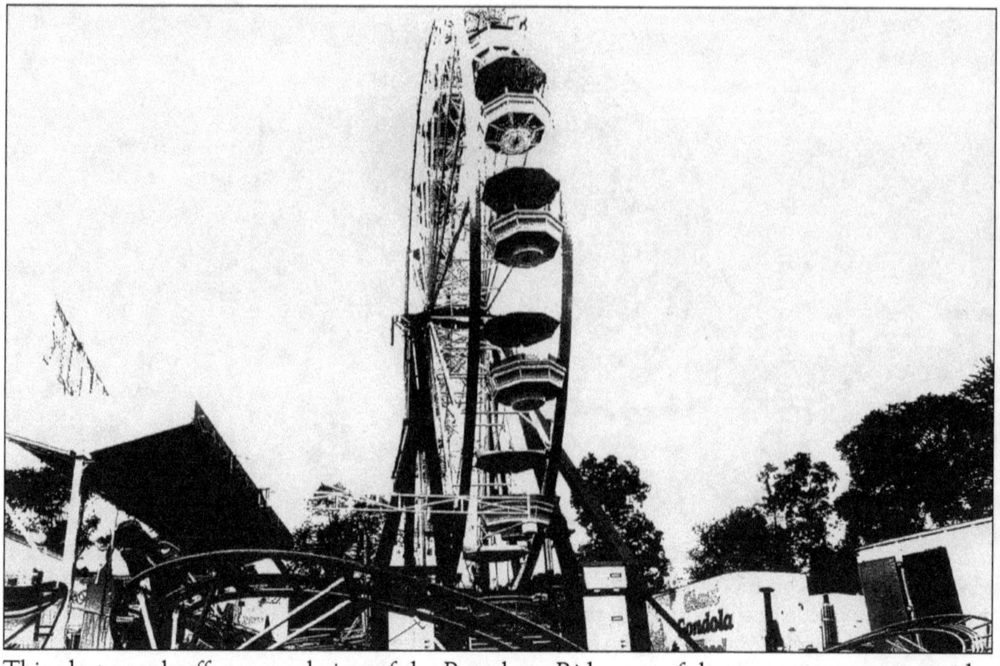

This photograph offers a good view of the Parachute Ride, one of the more contemporary rides at the Topsfield Fair midway.

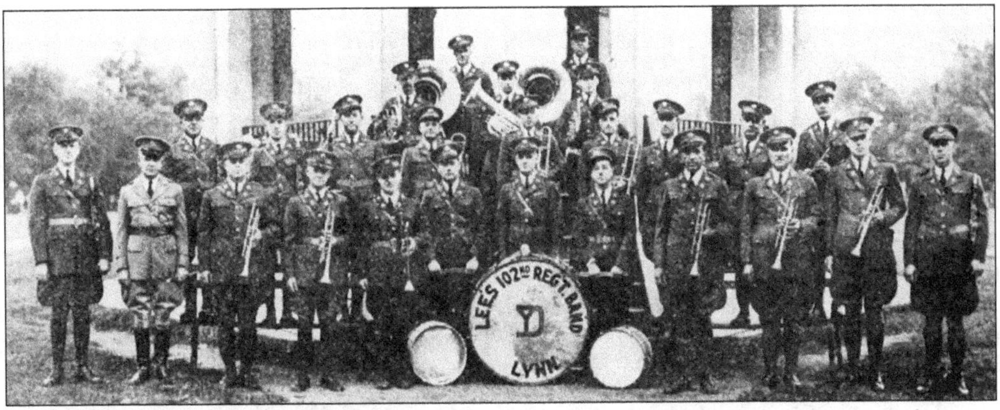

Lee's 102nd Regiment Band, a famous area band from Lynn, gave many concerts over the years at the Topsfield Fair in the 1930s.

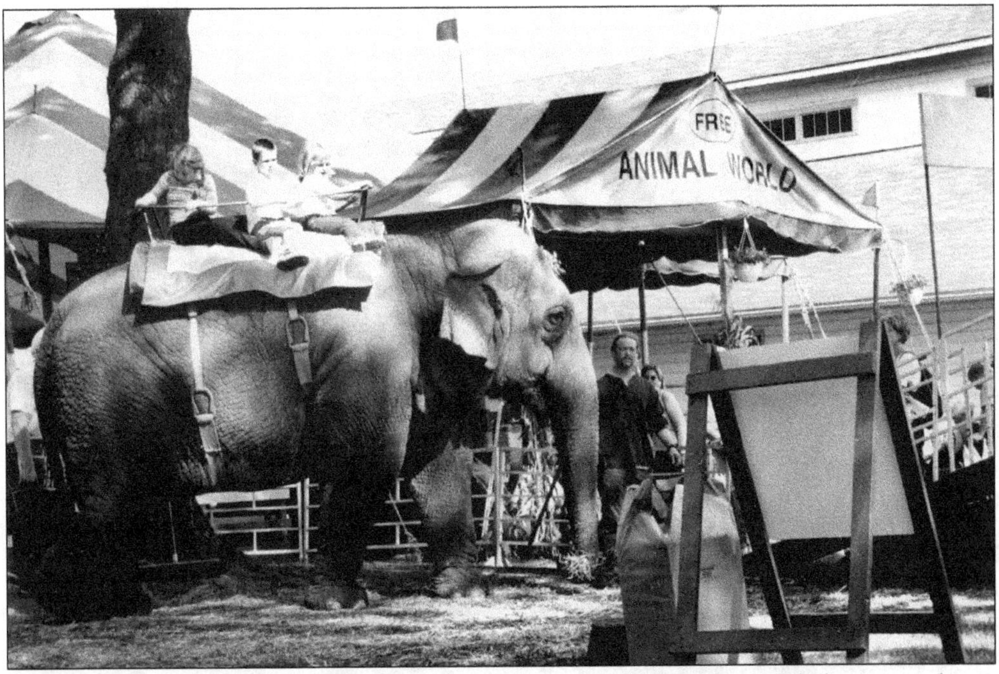

Located near the main gate in recent years, exotic animals from faraway habitats are shown to fairgoers. Some of them, such as this Asian elephant at the 2001 fair, even give rides to the delight of the children. The animals in the petting zoo are provided by Commerford and Sons from Goshen, Connecticut.

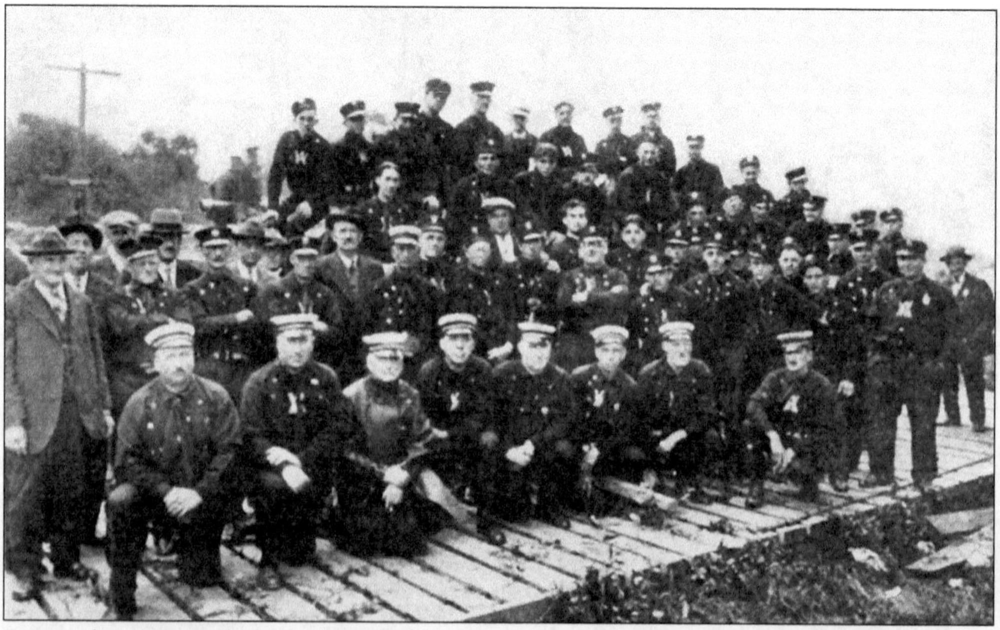

During the 1920s and 1930s, the Topsfield Fair hosted many firemen's musters. These firemen are assembled in 1929 on the Turnpike Crossing platform of the Boston and Maine Railroad on Route 1 just above the fairgrounds. Presumably, some of the fire companies brought their engines in to Topsfield on the train. The Boston and Maine provided special and regular train service for fairgoers, stopping at the Turnpike Crossing platform to let them detrain close to the fair's main gate.

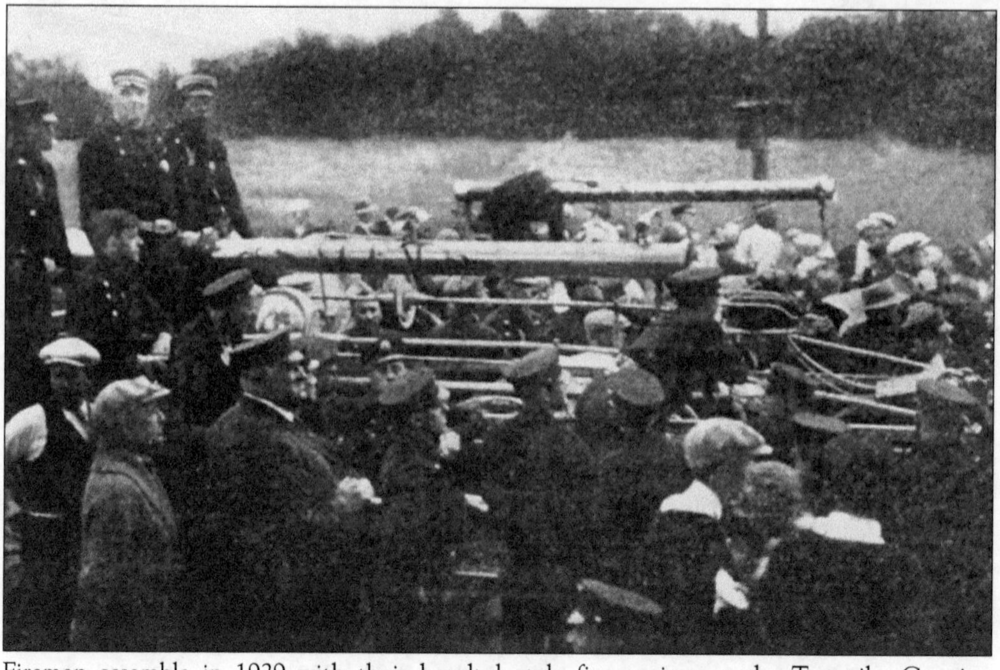

Firemen assemble in 1929 with their handtub-style fire engines at the Turnpike Crossing platform. The telephone box on the pole behind the firemen was probably used temporarily to communicate with the train from either the station in town or with the fairgrounds.

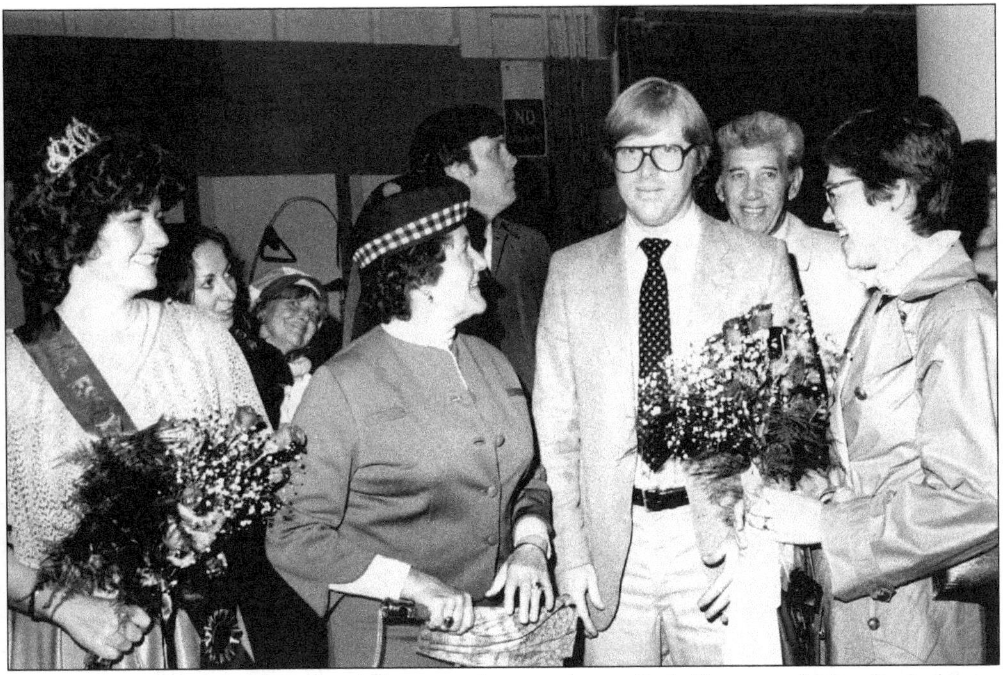

In a campaign swing to help his father win the presidency, Jack Carter and his wife, Judy, are given a tour of the flower show and the fair by Topsfield Fair president Nancy Begin. Mrs. Essex County and Sheila Barth are to the left, and behind Barth is Roberta Thompson. Behind Carter is Rep. Nicholas Mavroules.

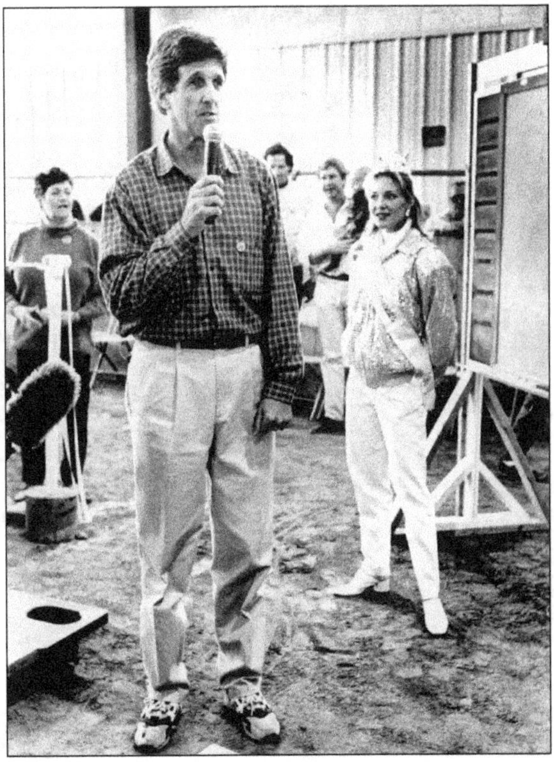

In 1996, Sen. John Kerry speaks to a large crowd at the All-New England Giant Pumpkin Contest weigh-off at the Topsfield Fairgrounds on Saturday as Miss Massachusetts, Lori Flick, looks on. Sally O'Maley, the fair's director of advertising and public relations, is standing in the background here in the livestock arena.

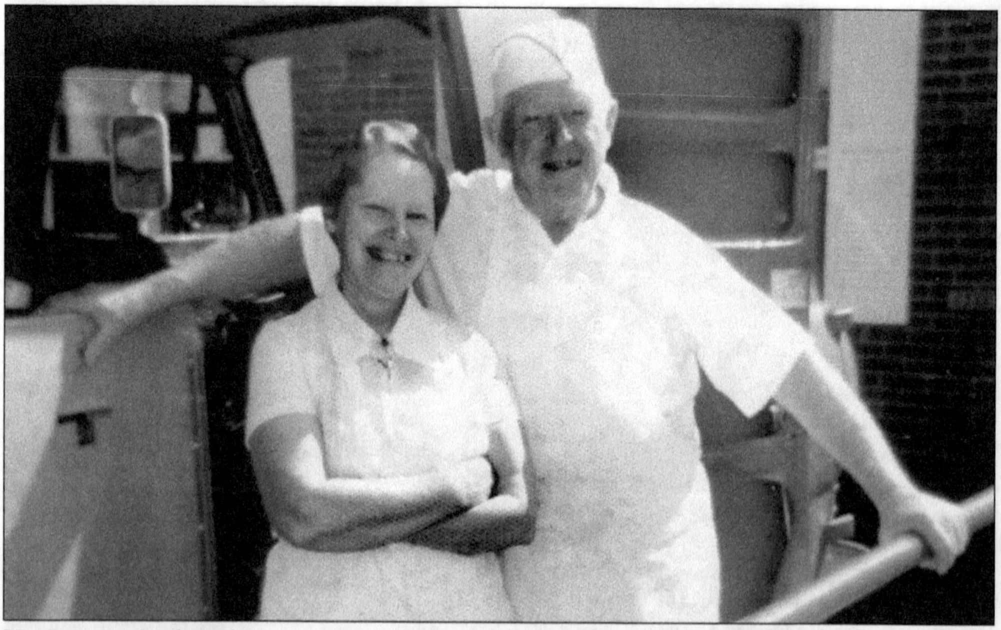

Here we see Ruth and Henry Nason of West Boxford loading their provisions truck to supply their barbequed-chicken stand at the Topsfield Fair, where by the mid-1950s, they had become longtime concessionaires and well known for their superior chicken barbeque dinners. (Photograph courtesy of Ruth Nason.)

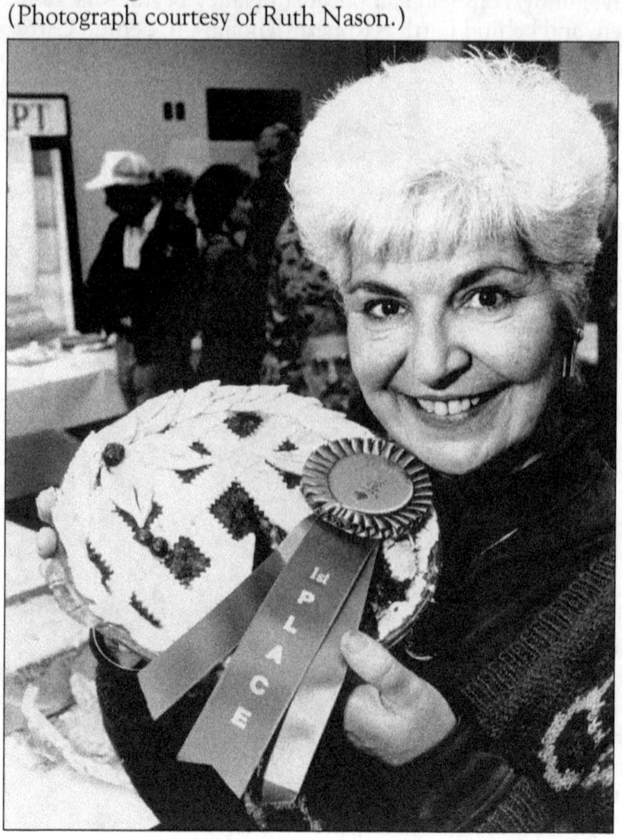

Ann Marshall of Hudson was the first prizewinner in the Crisco American Pie Celebration at the Topsfield Fair. The winning recipe was the "all-in-one hearty chicken pie."

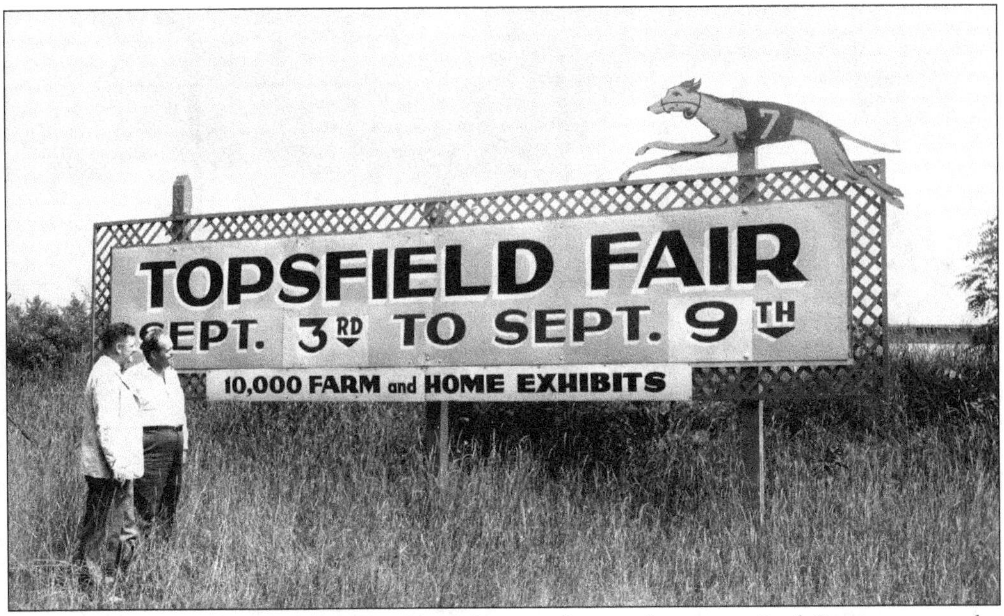

In this 1960s view, fair president Franklin Moore and general manager Paul Corson survey the large Topsfield Fair sign (with its upcoming fair dates) near the River Gate. Greyhound dog racing was a very popular event at the fair for decades until the meet was closed in the 1980s. Previously, sulky, flat, and steeplechase horse races, as well as motorcycle and auto races, were held here.

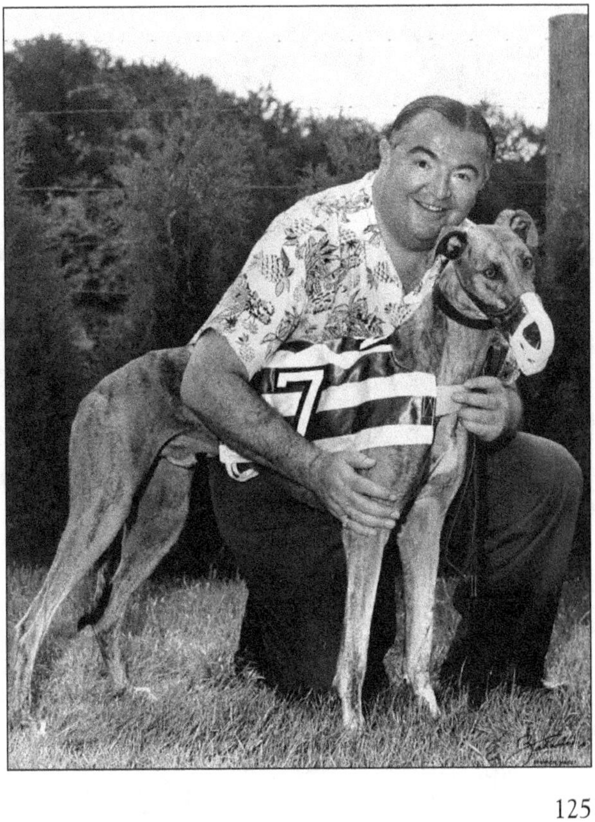

Tommy Maren was the director of the dog racing meets at the Topsfield Fair each season since he started there in 1950. In this 1971 photograph, he poses with one of his race-winning greyhounds.

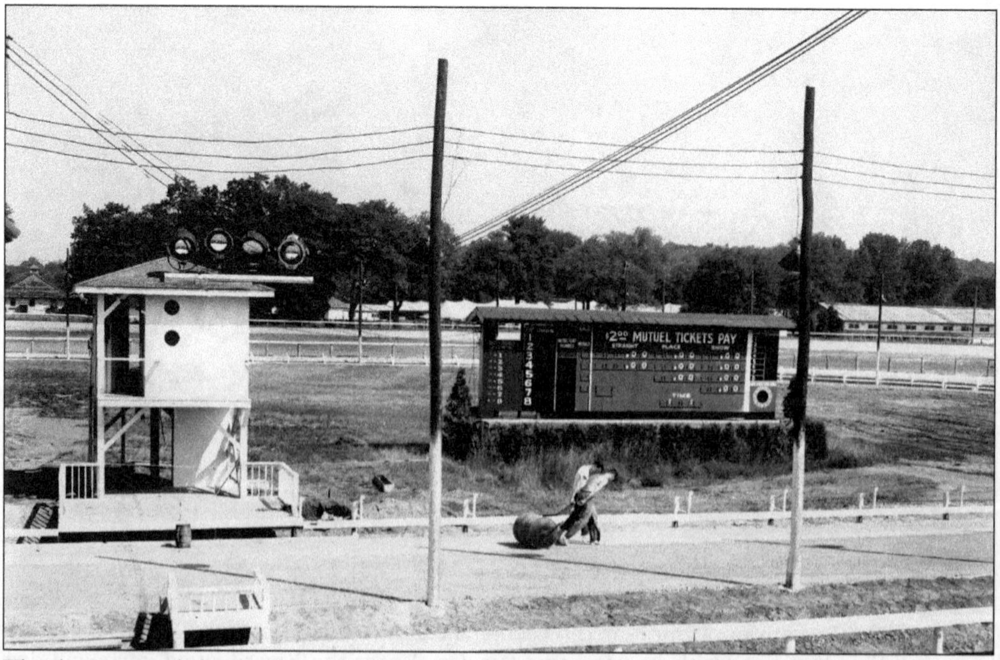

The dirt racetrack for the dog racing is being groomed by two young men as they roll the surface before the races commence later in the day in 1951. This is an excellent view of the tote board, judges' booth, and finish line as seen from the photographer's position.

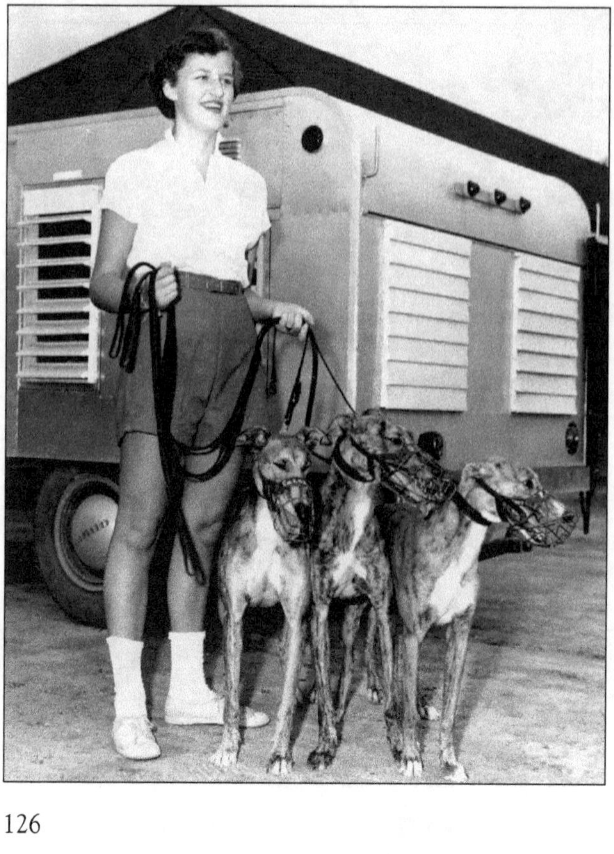

Exercising three greyhound pups that will appear on the speedy track at the Topsfield Fair nightly between September 1 and 6, 1952, is Susan Longo of Topsfield. These young speedsters will be pitted against leading dogs from Wonderland racetrack in Revere. There will be 10 races nightly at the floodlighted Topsfield Fair racing oval, and the racetrack fans will enjoy a new grandstand. Behind Longo is a dog trailer with individual compartments for each dog to travel in to the various race courses.

Augustus Means of Ipswich, the president of the fair from 1957 to 1960, tries to comfort a lost child, Michael Cameron of Arlington, who has become separated from his parents while touring the extensive fairgrounds. Lost children were taken to a small building near the main gate and cared for while the anxious parents were paged to come and retrieve their lost charges.

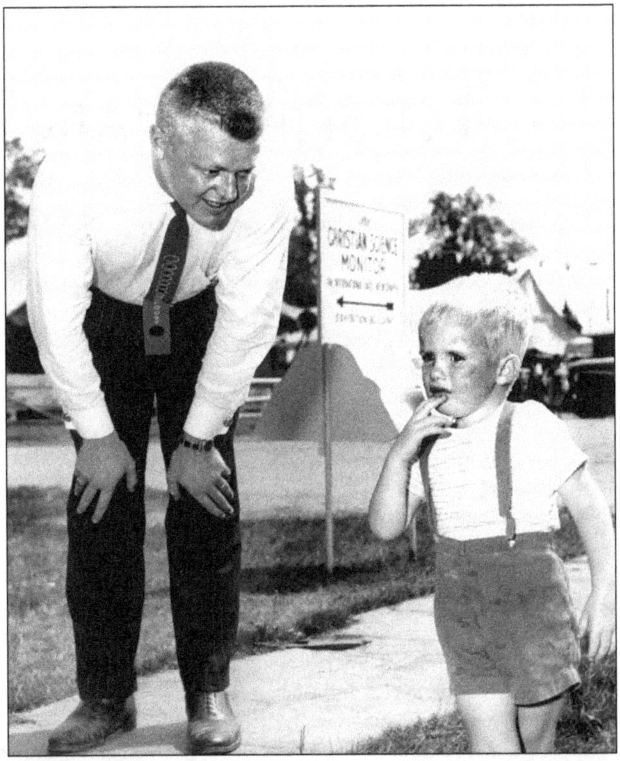

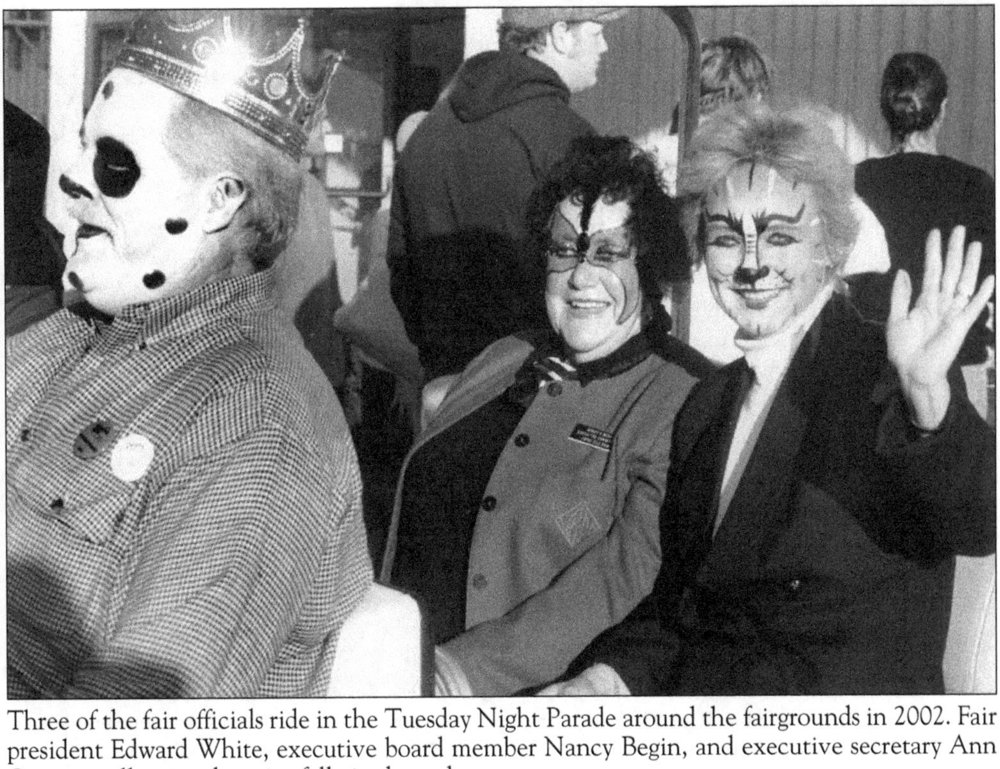

Three of the fair officials ride in the Tuesday Night Parade around the fairgrounds in 2002. Fair president Edward White, executive board member Nancy Begin, and executive secretary Ann Costanza all enjoy the crisp fall air along the way.

Topsfield Fair Timeline

1792: Massachusetts Society for the Promotion of Agriculture formed.

1810: Essex County Marino Sheep Company organized with a flock of 150 sheep.

1818: Essex Agricultural Society organized; Timothy Pickering of Salem, president.

1820: First cattle show of the Essex Agricultural Society; held for one day in Topsfield.

1828: The first cast shovel made in United States.

1831: First steel pitchfork made in United States.

1831: Essex County silk industry encouraged by Essex Agricultural Society; culture written up.

1832: First premiums given for horse teams; more farm machinery developed.

1835: Fruit and flowers first appear at the fair; apples widely promoted.

1844: Bees and honey appear at the fair at Ipswich.

1856: Site of present fairgrounds given by Dr. John Goodhue Treadwell.

1858: Ploughing matches with oxen most popular event at Danvers and well into later decades.

1870–1879: Improvement to farm wagons; one wagon (built by Daniels of Rowley) won first.

1893: First three-day fair held at Haverhill (three-day event continues for 17 years).

1895: Land purchased in Peabody; fair buildings and quarter-mile track built on 50 acres.

1895: Peabody becomes the site of the fair (remains the site for 15 years).

1910: Fair moved permanently to the Treadwell Farm (the present site), Topsfield.

1913: Essex County Agricultural School opens in Hathorne with 85 students.

1916: New buildings for horses, cattle, and swine opened.

1917: Essex County Agricultural Cooperative started by Essex Agricultural Society members.

1921: Exhibition hall, stable, grandstand, and dance pavilion built.

1923: Poultry building, cattle barn, horse barn, and racing stable built.

1923: Piggery, Women's Building (home arts), poultry building, and cattle barn built.

1924: Flower and fruit building (50 by 150 feet) built.

1924: Pony racing introduced on new racetrack; filling of swamp for more buildings.

1924: Secretary's office, ticket offices, and Women's Welfare and Red Cross Building constructed.

1924: Harness, flat horse, and steeplechase racing started; annual horse show started.

1924: Some 50,000 people attend fair; admission raised from 50¢ to 75¢.

1924: Boston and Maine Railroad regular and special trains stop at the Turnpike Crossing for the fair.

1927: "Topsfield Fair" (formerly "Essex County Fair") first used as the name of the fair.

1936: First full-time fair manager, Robert Trask of Winchester, hired.

1937: Midget car racing starts on a one-fifth-mile track in the infield of the horse track.

1938: Fair expands to five days.

1939: Goat building added.

1942: Ten percent of admissions given to army and navy relief.

1942: Free admission in exchange for 50 pounds of scrap metal or 20 pounds of scrap rubber.

1943–1945: Fair suspended due to the war.

1948: Paul Corson of Saugus becomes second full-time fair manager.

1950: Dean and Flynn contacted for midway show.

1953: Daniel Hurld's Old Essex County Barnyard opens on main oval (44 years at that location).

1963: The 4-H building opens; constructed by the Essex County 4-H Foundation.

1970s: Money Pole Climb begins.

1970: Royal Canadian Mounted Police (RCMP) starts Musical Ride at the fair.

1970: Trianon Amphitheater constructed.

1970: Essex Chapter FFA introduces hot cider to New England fairs at suggestion of RCMP.

1979: First opening-day fair parade in modern times.

1979: Draft horse show opens.

1980–1983: William Coolidge Center built; fair attendance exceeds half a million people.

1986: New administration building constructed.

1986: Mark Corson serves as general manager.

1987: Alvin Craig of Essex becomes the fourth fair general manager.

1990: New livestock show arena opens.

1993: Topsfield Fair celebrates 175th anniversary.

2001: Giant pumpkin exceeds 1,000 pounds.

2002: Pumpkin weighs 1,337 pounds; RCMP returns.

www.ingramcontent.com/pod-product-compliance
Lightning Source LLC
LaVergne TN
LVHW081933200926
848484LV00009B/2418